Reinventing the Automobile

Reinventing the Automobile

Personal Urban Mobility for the 21st Century

William J. Mitchell, Christopher E. Borroni-Bird, and Lawrence D. Burns

The MIT Press   Cambridge, Massachusetts   London, England

MIT Press books may be purchased at special quantity discounts for business or sales promotional use. For information, please email special_sales@mitpress.mit.edu or write to Special Sales Department, The MIT Press, 55 Hayward Street, Cambridge, MA 02142.

This book was set in Helvetica Neue LT Pro and Adobe Garamond Pro by the MIT Press. Printed and bound in the United States of America.

The views expressed in this book are those of the authors, and are not necessarily those of General Motors or the MIT Media Laboratory.

Library of Congress Cataloging-in-Publication Data

Mitchell, William J. (William John), 1944–.
Reinventing the automobile : personal urban mobility for the 21st century/William J. Mitchell, Christopher E. Borroni-Bird, and Lawrence D. Burns.
    p.   cm.
Includes bibliographical references and index.
ISBN 978-0-262-01382-6 (hardcover : alk. paper)
1. Electric automobiles—Technological innovations. 2. Intelligent transportation systems.   3. Transportation, Automotive.   4. Urban transportation.   I. Borroni-Bird, Chris.   II. Burns, Lawrence D.   III. Title.
TL220.M58   2010
629.2—dc22
                                              2009024970

10   9   8   7   6   5   4   3

Contents

Preface

Imagine driving around your city in one of the vehicles shown on the cover—while connecting to your social network and favorite news and entertainment sources, using your time efficiently, and expending only renewable energy. This book presents four big ideas that will make this possible. It weaves them together into a comprehensive vision for the future of automobiles, personal mobility systems, and the cities they serve.

The first idea is to transform the DNA—that is, the underlying design principles—of vehicles. The DNA of today's cars and trucks depends on petroleum for energy, on the internal combustion engine for power, and on manual control and independent, stand-alone operation. The new automotive DNA is based on electric-drive and wireless communications. It will allow future vehicles to be lighter and cleaner, drive themselves when necessary, avoid crashes, and be fun and fashionable.

The second idea is the Mobility Internet. This is a logical development from its predecessors—the computer Internet, the cell-phone Internet, and the "Internet of things" enabled by electronic tags and sensors. It will enable vehicles to collect, process, and share enormous amounts of data so that traffic can be managed and travel times can be reduced and made more predictable. It will also permit drivers to remain seamlessly connected to their social networks.

The third idea is to integrate electric-drive vehicles with smart electric grids that use clean, renewable energy sources—particularly solar, wind, hydro, and geothermal—together with dynamic electricity pricing. This not only provides clean energy to vehicles, but also enables grids to operate more efficiently and to make more effective use of renewables. By taking advantage of the electricity storage capacity of electric-drive vehicles and employing price signals to regulate demand and mitigate the effects of the intermittent supply characteristic of many renewable sources, smart grids can keep electricity supply and demand in optimal balance.

The fourth idea is to provide real-time control capabilities for urban mobility and energy systems. This is accomplished by establishing dynamically priced markets not only for electricity, but also for road space, parking space, and in some contexts shared-use vehicles. The wireless connectivity and onboard intelligence of the automobiles that we propose enables them to respond appropriately to the price signals within these markets. This provides an effective way to balance supply and demand, relieve road and parking space congestion, and increase the utilization rates of available vehicles.

Why haven't these ideas been pursued in concert in cities before? Many of their elements, after all, aren't new. The answer is that the enabling technologies not only had to develop, but also had to converge before they could become effective. They have now done so. This creates an opportunity to reinvent automobiles and personal urban mobility systems fundamentally (not just improve them incrementally), which is what's needed to meet the urgent sustainability challenges we face.

This fundamental reinvention will enable the creation of automobiles that weigh less than a thousand pounds, are less than a hundred inches long, and do better than 200 miles per gallon of gasoline on an energy-equivalent basis. They can provide safe, convenient personal urban mobility at about one-quarter the total cost per mile of today's cars, take up approximately one-fifth of the space currently needed in cities for parking, significantly improve the throughputs of streets and roads, and eliminate carbon emissions.

Reinvented automobiles will have the most profound effects in cities and towns, where more than half of the world's people now live, and where an estimated 80 percent of the world's wealth will be concentrated by 2030. Cities continue to attract population because they provide access to resources and opportunities, but they are also where the energy, environmental, safety, congestion, and spatial externalities of today's cars are most strongly amplified. Reinventing the automobile will create the opportunity for cities to become more livable, equitable, and sustainable.

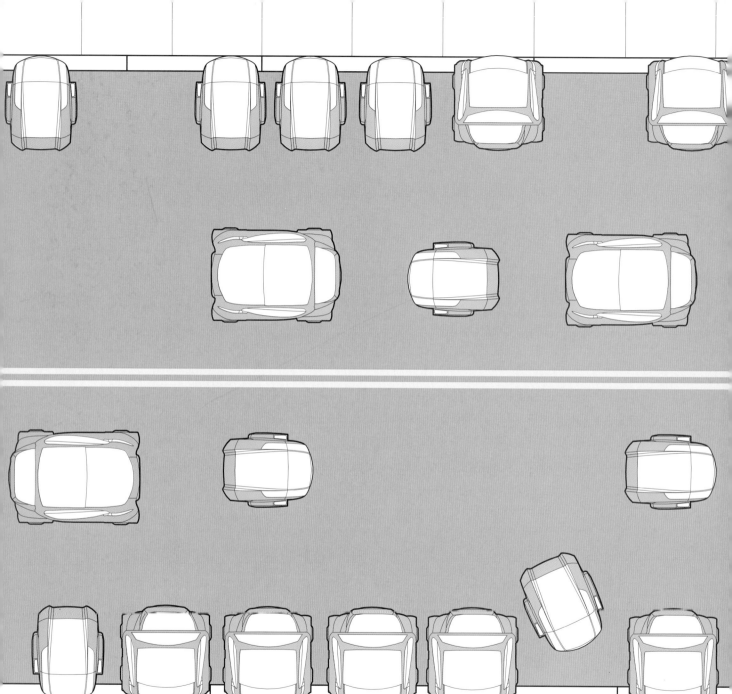

Introduction

For a century, the automobile has offered affordable freedom of movement within cities—the places where most of the world's people now live, work, play, and pursue their social and cultural lives. It provides access to all of the benefits that cities have to offer; it is an object of desire; and it plays a crucial role in the U.S. and other economies. But it now requires radical reinvention.

Through a complex coevolutionary process—involving interdependencies among vehicle engineering and design solutions, energy supply systems, street and road infrastructures, urban land use patterns, economic incentives, and government policies—the automobile has become part of the urgent problem that cities now face. Cities currently consume too much of the Earth's nonrenewable resources to remain viable and livable in the long term. Their supply lines are insecure and vulnerable to disruption. They are too congested with parked and moving vehicles to be safe, convenient, and pleasant. And they produce more waste—including the greenhouse gases associated with global warming—than the Earth's natural systems can absorb without undergoing unacceptable levels of damage.

This book argues that a reinvented automobile can become a powerful part of the solution to these problems. While maintaining and even enhancing current levels of personal mobility within cities, the new kinds of automobiles and personal urban mobility systems that we'll describe promise to reduce the overall energy and materials requirements of mobility systems; facilitate a significant shift from nonrenewable energy sources to clean, renewable ones; eliminate tailpipe emissions; enhance energy security; and generally improve the quality of urban life.

These automobiles are also designed to have high consumer appeal—to be fun, fashionable, and affordable. This is crucial: It is only through very high-volume consumer acceptance that reinvented automobiles and mobility systems will make the large-scale contributions to urban sustainability that we need, create exciting new opportunities for the automobile industry, and help to establish a clean, green economy for the coming decades.

The Need for Sustainable Personal Mobility

Automobiles respond to our desire to move about and interact. Ever since our ancestors walked out of Africa, personal mobility has been recognized as a basic human need. The transportation of people and objects and the creation of systems for moving freely from one place to another have been a part of the human story from prehistory.

From clans to cities, caves to skyscrapers, walking to riding, and sandals to cars, we have a rich history of finding ways to grow our population and wealth by increasing our mobility and our access to various resources. The invention of the wheel enabled hand-pulled and animal-drawn carts, and the domestication of horses extended the range of travelers. Horses remained a leading source of transpor-

tation power until they were supplanted, a hundred years ago, by mass-produced automobiles.

While automobile transportation has dramatically enhanced our personal mobility and helped us realize our aspirations for growth and prosperity, it has also created troublesome side effects. The freedom and prosperity benefits have been substantial, including greater access to jobs, goods, and services, convenient and safer personal travel, and the ability to go where we want, when we want, while carrying the things that we need. At the same time, however, the side effects have also been significant and growing. In our pursuit of personal mobility, we have damaged our environment, consumed our natural resources, wasted our time in traffic, harmed each other in collisions, and created disparities between the haves and have-nots. The extrapolation of these side effects raises increasingly pressing questions about the sustainability of today's automobile transportation system. Fortunately, rapidly maturing and converging technologies promise to reduce, and in some cases eliminate, these negative effects while further enhancing our freedoms. This book provides a comprehensive vision for the future of automobiles and personal urban mobility based on this promise.

The numbers are staggering. Over 6.7 billion people reside on Earth, with more than half of us now living in urban areas. This includes 26 cities with populations exceeding ten million people.[1] We own 850 million cars and trucks, nearly all powered by internal combustion engines and energized with petroleum. Parked end to end, these vehicles would circle our planet nearly one hundred times—yet this represents a motor vehicle for just one out of every eight of us.

In the United States, 85 percent of personal travel today is by automobile. Americans drive three trillion miles a year, on four million miles of roads, consuming 180 billion gallons of fuel each year dispensed from 170,000 service stations.[2] Furthermore, we can expect significant increases in the number of cars being sold in emerging markets. With a sales growth rate of 3 percent per year, China's vehicle population is projected to surpass that of the United States by about 2030.[3] And as India's economy expands, it is poised to follow in China's footsteps.

Worldwide, we consume 18 million barrels of oil each day driving cars. Our vehicles emit 2.7 billion tons of carbon dioxide each year.[4] Roadway collisions claim 1.2 million lives each year.[5] And, in dense city centers, average urban speeds today can be well under 10 miles per hour.[6]

Have we reached the point where we now must seriously consider trading off the personal mobility and economic prosperity enabled by automobile transportation to mitigate its negative side effects? Or can we take advantage of converging twenty-first-century technologies and fresh design approaches to diminish these side effects sufficiently while preserving and enhancing our freedom to move about and interact? This book concludes the latter. It weaves together four big ideas that, when combined, hold the promise of sustainable automobility, even for dense megacities. Though some of the elements of these ideas are not new, we believe that it is now necessary—and entirely feasible—to develop and combine them in a radically new way.

Four Ideas: A Summary

The first idea, detailed in chapter 2, is to adopt a new automotive DNA that transforms the design principles that currently underlie automobiles. As summarized in figure 1.1, today's cars and trucks are primarily mechanically driven, powered by internal combustion engines, energized by petroleum, controlled mechanically, and operated as stand-alone devices. In fact, they have essentially the same "genetic

Current DNA	New DNA
Mechanically driven	Electrically driven
Powered by internal combustion engine	Powered by electric motors
Energized by petroleum	Energized by electricity and hydrogen
Mechanically controlled	Electronically controlled
Stand-alone operation	Intelligent and interconnected

Figure 1.1
The new automotive DNA.

makeup" as automobiles pioneered by Karl Benz, Ransom Olds, and Henry Ford over a century ago.

The new automotive DNA is created through the marriage of electric-drive and "connected" vehicle technologies. It is based purely on electric-drive, using electric motors for power, electricity (and its close cousin, hydrogen) for fuel, and electronics for controls. Electric-drive vehicles include battery electrics, extended-range electrics, and fuel-cell electrics. All three of these vehicle types have important roles to play in our future and differ from now-familiar hybrid electric vehicles, which add batteries and electric motors to improve the efficiency of today's mechanically driven cars.

The new automotive DNA also allows vehicles to communicate wirelessly with each other and with roadway infrastructure and roadside activities. When combined with GPS (Global Positioning System) technology and information-rich digital maps, "smart" cars will know precisely where they are relative to everything around them. Even with today's technology, vehicle-to-vehicle (V2V) communications and GPS can allow us to determine the proximity of two vehicles to within a meter and predict where these vehicles will be during the next twenty milliseconds. Taking advantage of such capabilities, connected vehicle technology will enable cars that can drive themselves and avoid crashes. The resulting reduction in crash protection requirements means that cars can become lighter, making them more conducive to electric drive, and thereby encouraging the use of renewable sources of energy for personal transportation. It also means that cars

can be even more fun to drive and can provide more freedom of expression and personalization.

The second idea, the Mobility Internet, is discussed in chapter 3. The Mobility Internet will do for vehicles what the Internet has done for computers. It will enable vehicles to share enormous amounts of real-time, location-specific data so that traffic can be managed optimally and travel times can be reduced and made more predictable. Just as today's Internet servers manage extraordinary amounts of e-mail traffic, the Mobility Internet servers will manage vast amounts of vehicle traffic. This will integrate vehicles into the emerging "Internet of things."[7] Automobiles will become nodes in mobile networks.

The Mobility Internet will also permit drivers to share information and remain seamlessly connected to their personal, social, and business networks. Nondriving passengers will be able to do this soon. And, when automobiles begin to drive autonomously, even those in charge of automobiles will be able to safely use their travel time as they please, because there will no longer be the "distraction of driving."

The combination of the automobile's new DNA and the Mobility Internet, when applied within cities and towns, will enable us to reinvent personal urban mobility systems for the twenty-first century. Vehicles designed for city use will have dramatically smaller spatial and carbon footprints and will be considerably less expensive to own and operate. Later, we will introduce two new personal mobility concepts based on the new automotive DNA. These concepts stem from work done at MIT and General Motors and illustrate just a couple of the many design and styling opportunities made possible when electric-drive vehicles are connected and enabled to avoid crashes and drive autonomously. They are extremely mass, space, and energy efficient. They provide all-weather protection, are comfortable, and allow their occupants to socialize, both physically and virtually. They are works in progress rather than fully designed and engineered products, but they clearly demonstrate the design directions that are possible. They are discussed in chapter 4.

The third idea is smart, clean energy, discussed in chapters 5 through 7. This results from combining electric-drive vehicles with energy-efficient buildings and smart utility grids to create distributed, responsive energy systems. These systems will support the utilization of diverse and renewable (but intermittent) sources of electricity. In addition, because electricity and hydrogen are interchangeable and hydrogen can store energy more densely than batteries, smart energy systems will enable the optimal mix of batteries and fuel cells to facilitate both stationary and vehicle uses of electricity. This includes the potential to efficiently distribute small amounts of energy precisely when and where they are needed.

The final idea is to develop electronically managed, dynamically priced markets (discussed in chapter 8) for electricity, roads, parking, and vehicles. These markets are underdeveloped today, but stationary and mobile connectivity can help realize their potential. They will depend on ubiquitous metering and sensing, make use of powerful computational back-ends, provide price signals and incentives that regulate supply and demand, and motivate sustainable activity patterns within cities.

The Combination of Transformative Ideas

Taken individually, each of these four ideas offers significant individual and societal benefits. Each can be implemented more or less separately. When pursued together, though, they will have their greatest impact. They have the potential to radically transform personal mobility in cities. To illustrate their power in combination, chapter 9 explores their combined effect on cities, where we can expect most of the world's population, together with 80 percent of the world's wealth, to be concentrated by 2030 (according to the United Nations). Cities will continue to attract population because they provide the greatest access to resources and opportunities. However, they are also the places where the energy, environment, safety, congestion, and access-inequality side effects of today's automobiles are most strongly amplified.

When effectively combined, the ideas behind this reinvention promise to enhance our freedoms and stimulate economic growth and prosperity while eliminating many, if not all, of the negative side effects of today's automobile transportation system. Figure 1.2 summarizes this opportunity.

The enabling technologies underlying these ideas have only recently matured and begun to converge, and this has fulfilled a necessary condition for large-scale, comprehensive application. Feasibility now

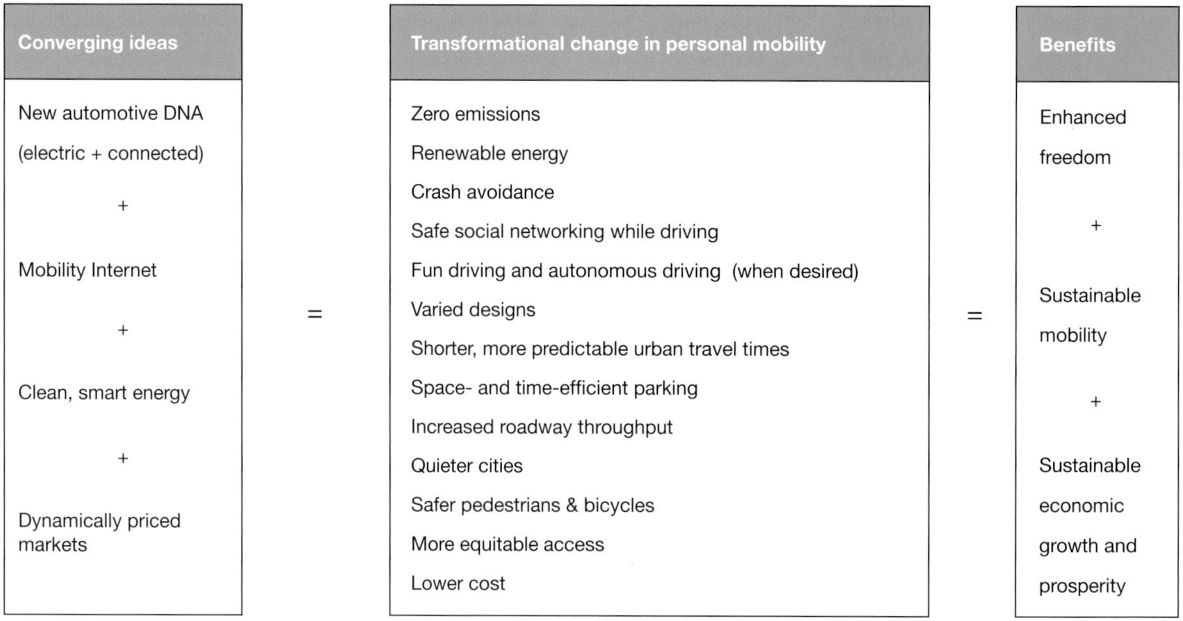

Converging ideas		Transformational change in personal mobility		Benefits
New automotive DNA (electric + connected)		Zero emissions		Enhanced freedom
		Renewable energy		
+		Crash avoidance		+
		Safe social networking while driving		
Mobility Internet	=	Fun driving and autonomous driving (when desired)	=	Sustainable mobility
		Varied designs		
+		Shorter, more predictable urban travel times		+
		Space- and time-efficient parking		
Clean, smart energy		Increased roadway throughput		Sustainable economic growth and prosperity
		Quieter cities		
+		Safer pedestrians & bicycles		
		More equitable access		
Dynamically priced markets		Lower cost		

Figure 1.2
The whole is greater than the sum of its parts.

meets need. We have before us an emerging opportunity to reinvent the automobile and personal mobility systems fundamentally, not just improve them incrementally, and in doing so to meet the pressing sustainability challenges we face.

Implementation

We close in chapter 10 by discussing what must be done to realize this vision. The successful development of the Internet has demonstrated that large-scale networks and the necessary integration among diverse products and businesses can be realized. And there are many useful lessons to be learned from this example.

Several of the necessary technology enablers have already been tested in market "footholds." Others need foothold tests, and all need to mature to reach market "tipping points." Because of the codependencies among automobiles, energy infrastructure, communications infrastructure, and governments, it will be necessary to align incentives, form coalitions of stakeholders, and build a broad consensus around a common vision of the future of personal urban mobility. This common vision must be constructed around a "system-of-systems" framework, with widely accepted standards enabling interfaces among systems, and the increase in value of a system for every user as it grows (sometimes known as its "positive network externalities") motivating investment and powering rapid system growth. New forms of public–private partnership, as well as new business models, will be needed to realize the public and private value embedded in the vision and to share its risks and rewards. Finally, we must engage the imaginations of citizens, vividly demonstrate what is at stake and what is possible, and stimulate political support and consumer demand.

Among the first steps will be the development of imaginative, carefully conceived pilot projects. These must be at sufficient scale, and at sufficient levels of investment, to enable the integration of the key ideas. These projects will demonstrate the feasibility of reinvented automobiles and personal urban mobility systems, illustrate the daily operational and experiential realities, provide opportunities for necessary experimentation and testing, enable large-scale data collection and analysis, and provide the experience and learning needed to move to larger scales.

The Road Ahead

Getting to where we need to be will require a lengthy and difficult journey. (Remember, the initiation of what became the Internet took place back in the late 1960s.) But the outcomes can include much more livable and sustainable cities, economic growth based on clean, green technologies, and prosperity and freedom for future generations.

The New DNA of the Automobile

The shape of today's automobiles is derived from the placement of the engine (under the hood) and was a natural evolution from its predecessor, the horse and carriage. This arrangement enabled the transition from real horse power in the nineteenth century to mechanical power in the twentieth century.

In the twenty-first century, we will see a further transition from mechanical power to electric power. As vehicles are increasingly propelled by electric motors and energized by electricity, a new freedom to innovate in automobile design is emerging. The electric-powered skateboard, shown in figure 2.1, was a pioneering demonstration of this.

When provided with sufficient wireless connectivity and intelligence, future vehicles will drive themselves and have advanced crash-avoidance capabilities. This is not science fiction, but convincingly demonstrated technology that awaits reduction in bulk and cost and being put to good use. It will provide additional freedom to simplify and reimagine vehicle structures and interiors. The emergence of a new automobile DNA, derived from these changes, promises a renaissance in vehicle design. It will open up for exploration spaces of design possibilities that have never before been seriously considered.

The Evolution of the Automobile and Its DNA

Although motorized conveyances were envisioned by Leonardo da Vinci in the fifteenth century, the

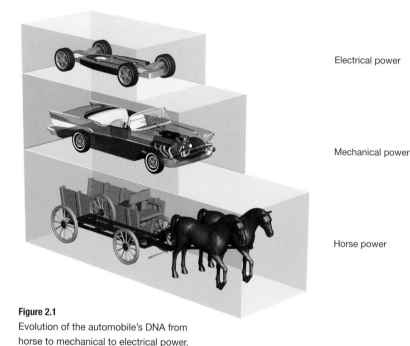

Electrical power

Mechanical power

Horse power

Figure 2.1
Evolution of the automobile's DNA from horse to mechanical to electrical power.

first truly self-propelled vehicle was not developed until 1769, when the French engineer Nicolas-Joseph Cugnot mated a steam engine with an artillery wagon for the French army. A year later, he built the first steam car, a three-wheeled vehicle that carried four passengers. It was a tentative beginning, and yet, only fifty years later, engineers, mechanics, and tinkerers around the world were working on "horseless carriages."

Today, the design of the automobile seems inevitable, but in the beginning it was anything but certain. There was a wide variety of power plants—including steam engines, electric motors and batteries, and internal combustion engines—and it was not even entirely clear whether vehicles should have two, three, or four wheels.

The steam engine was a popular propulsion choice for early developers because it was a well-understood technology that had been used in industrial applications for more than a hundred years. The first "steamers," however, had some significant negatives. They took 30–45 minutes to go from start-up to "steam-up" and could only be driven about 30 miles before having to take on more water. Although flash boilers and exhaust condensers eventually overcame these disadvantages, these innovations came too late to save steam-powered automobiles.[1]

Electric motors and batteries also gained early favor. The first electric carriages were built in the 1830s by Robert Anderson in Scotland and Sibrandus Stratingh and Christopher Becker in the Netherlands. Breakthroughs by Gaston Plante and Camille Faure increased battery energy storage capacity, which led to the commercialization of battery-electric cars in France and Great Britain beginning in the 1880s and in the United States in the 1890s. Battery-electric vehicles were quiet, clean, and simple to operate, but their batteries took a long time to recharge, were expensive to replace, and had limited range. As a result, battery-electric cars, like the steamers, were mainly "city" vehicles.[2]

The internal combustion engine (ICE) was a third propulsion choice, and it, too, had both benefits and shortcomings. Early proponents believed that these engines offered outstanding power, speed, and range potential, but they were also complex, noisy, and dirty. They required a dangerous hand-crank to start, needed a gear system to transmit power from the engine to the drivetrain, used combustible fuel, and produced foul-smelling exhaust. Despite these problems, inventors pressed on. In 1885, Karl Benz designed, built, and patented a three-wheeled vehicle in Germany that today is recognized as the first gasoline-fueled motor car, while similar vehicles were being developed across Europe and the United States.[3]

Given the drawbacks of the first automobiles, at the turn of the twentieth century, the jury was still out on which propulsion system was the right solution for motorized transportation. The internal combustion engine was gaining support in engineering circles, but most of the vehicles manufactured in the United States were still battery electric or steam.

And yet, a remarkable convergence of technology, manufacturing know-how, business skills, energy supply, roadway development, public policy, and consumer demand would soon make the internal combustion engine the predominant automotive power plant, triggering an economic growth dynamic that ultimately helped enable enormous prosperity.

An important technology enabler was Charles Kettering's invention of the electric self-starter, which eliminated the hand-crank and made ICE vehicles easier to start, safer, and more accessible.[4] Another critical event was the improvement in America's roadways to the point where vehicles could carry people and cargo over greater distances. In addition, the discovery of large reservoirs of crude oil in Oklahoma and Texas, coupled with petroleum's favorable energy density, made petroleum more readily available and gasoline less expensive than other fuels. Another key enabler was Ransom Olds's and Henry Ford's implementation of the automotive assembly line and low-cost mass production, which made ICE-powered automobiles affordable to large numbers of people.[5] Finally, Ford's introduction in 1914 of a $5-per-day pay scale enabled the production workers to purchase the products they built and helped set the stage for the growth of the middle class.

The rest is familiar history. The internal combustion engine became the powerhouse that made the automobile the dominant mode of personal travel. It drove the growth of new industries devoted to automotive manufacture, fueling, and road building, which ultimately culminated in "a vehicle for every purse and purpose" and the interstate highway system. These, in turn, enabled modern suburbia, opened up a new American way of life, and powered a century of economic prosperity. As the foundation, the ICE may not have been the ideal option, but it was the best available solution for its time.

The growth in automobile ownership that began in the early decades of the twentieth century was fed by the universal desire for increased freedom to pursue and fulfill our personal aspirations and to go where we want, when we want. It was fertilized by the synergy among automobiles, roads, and fuel stations.

At least 14 million U.S. jobs today are tied directly or indirectly to producing and servicing automobiles, building and maintaining roads, supplying fuel, and governing our infrastructure. This makes personal transportation a major economic engine for the United States. In fact, nearly 20 percent of the nation's retail sales are automobiles, and the vehicle industry represents 20 percent of our manufacturing sector's output. The cost to individual consumers is also not trivial—around 50

Cost categories	Annual cost (midsize sedan)
Gasoline	$ 1,851
Maintenance	$ 701
Tires	$ 127
Operating costs	$ 2,679
Insurance	$ 907
License, registration, and taxes	$ 562
Depreciation	$ 3,355
Financing (10% down, 5 year loan at 6%)	$ 770
Ownership costs	$ 5,594
Total costs	$ 8,273

Figure 2.2
The cost of driving a midsize sedan 15,000 miles in 2008 equates to around 55 cents per mile (assuming gasoline price of $2.94/gallon).

cents per mile for a midsize sedan like a Chevrolet Malibu, Toyota Camry, or Honda Accord, according to AAA's Annual Cost of Driving study for 2008 (figure 2.2).[6] And this ignores the cost of parking.

Emerging Problems and Opportunities

Over the last hundred years, successive technology breakthroughs have made our vehicles more powerful, easier to drive and control, safer, more energy efficient, and more environmentally friendly. While the improvements have been dramatic, the innovations themselves have been largely evolutionary. In fact, the basic DNA of the automobile has not really changed all that much. Just like the first mass-produced automobiles, our vehicles continue to be powered by the internal combustion engine, energized by petroleum, driven and controlled mechanically, and operated as stand-alone devices.

Today, though, there is a growing realization that the 120-year-old foundational DNA of the automobile is no longer tenable. While the scale and scope of the automobile transportation system and the freedom and prosperity it provides are impressive, we must seriously question whether the present system can be sustained into an increasingly crowded, resource-constrained future—particularly in urban areas. Today's vehicles are typically designed to meet almost all conceivable needs for transporting people and cargo over both short and long distances, but this amazing flexibility drives considerable cost and inefficiency (mass, space, and energy) into the vehicle and plays a large role in creating the difficulties of dependence on imported oil; air pollution and greenhouse gas emissions; fatalities and injuries to drivers, passengers, pedestrians, and other road users; traffic congestion; inefficient land use; and access disadvantages for those without cars. These difficulties are amplified in cities where congestion is worst, parking space is limited, and pedestrian and cyclist accidents are a significant issue. In this particular environment, where most of the vehicles are used and operated, the automobile, as currently configured, is overengineered.

Today, a typical automobile weighs 20 times as much as its driver, can travel over 300 miles without refueling, is able to attain speeds well over 100 miles per hour, requires more than 100 square feet for parking, and is parked more than 90 percent of the time. Unless an automobile is being used to haul large numbers of people or heavy cargo most of the time, this is significantly beyond what is necessary for providing safe, convenient, efficient personal mobility within cities. By designing to specifications that more closely match the requirements of personal urban mobility we can reduce mass and material use, save space, and save energy.

Fortunately, just as it was a century ago, we have within our grasp a remarkable transformation of personal urban mobility—one that has the potential to catapult our freedom of movement to a new plateau, create even greater economic growth and prosperity, and do so sustainably in terms of energy, the environment, safety, and congestion. We do not need to invent anything for this to happen. The task is to act with a collective will to accelerate the implementation of what is already technically doable and holds the promise to be viable commercially.

The first step is to create a new automotive DNA. With this new DNA, our vehicles will be electric-drive, fueled by electricity and hydrogen, electronically controlled, and will function as nodes in a connected transportation network. They will be able to communicate wirelessly with each other, with the roadway infrastructure, and with roadside activities. They will also know precisely where they are in relation to other vehicles and the roadway, and will be able to drive themselves and avoid crashes and traffic tie-ups.

This new automotive DNA goes beyond the re-invention of the automobile around electrification, as described by Lawrence Burns, J. Byron McCormick, and Christopher Borroni-Bird in *Scientific American* in October 2002 ("Vehicle of Change"). Electrification will enable vehicles to be more energy efficient, cleaner, safer, and more fun to drive, but we will also need connectivity among vehicles to improve the coordination of vehicles on the roads, to reduce accidents and to manage congestion. Allowing vehicles to operate more effectively as a system will provide additional improvements in energy efficiency and air quality. It is the combination of electrification and connectivity that promises to revolutionize personal mobility. Importantly, the technology needed for this is already available.

This emerging transformation of the automobile's DNA will be similar to what has already occurred with computers and communications. Today's vehicles are analogous to the stand-alone desktop personal computers of the 1980s. Tomorrow's vehicles will be analogous to handheld devices with Internet connectivity—smaller but, in many important respects, more capable and flexible.

The New DNA (Part 1): Electrification

The new DNA for the automobile will have an electric-drive foundation. Conventional automobiles are mechanically driven. Hybrid electric vehicles, even plug-in hybrids, combine a conventional me-

chanical drive with electric-drive, and this added complexity introduces extra cost and mass into the vehicle. Like today's conventional gasoline engine–powered automobiles, electric-drive vehicles will have a single propulsion system, but one that is driven electrically with a motor, rather than mechanically with an engine and transmission.

The growing interest in electric-drive vehicles has been motivated by recent improvements in lithium-ion batteries and hydrogen fuel cells. New battery chemistries have enabled more stable battery cells that have higher energy and power density. Today, these cells are being developed into safe battery packs that have both the high energy needed for increased battery-electric range and the high power required for acceleration in a package size that fits a compact car.

With these advancements, car companies and battery developers are partnering on a range of battery-electric vehicles. Many companies, not just the traditional automakers, have recently announced electric-drive vehicle demonstration and production programs (some are shown in figure 2.3). These vehicles will store and create electricity on board and will be driven purely electrically.

Examples of these new kinds of vehicles include:

- Battery-electric vehicles (BEVs) like the Think City, Tesla Roadster, BMW MINI E, Daimler Smart Fortwo, and Mitsubishi iMIEV, which have reported driving ranges varying between 50 and 240 miles.
- Plug-in hybrid versions of the Toyota Prius and Ford Escape, which may soon be on the market and can be operated for 10–40 miles on the battery alone.

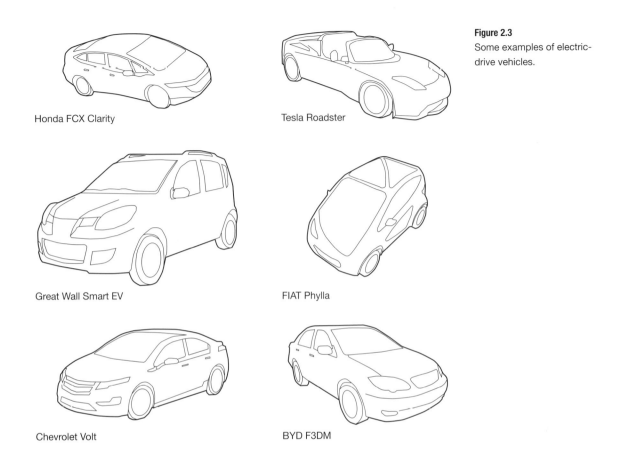

Figure 2.3
Some examples of electric-drive vehicles.

Honda FCX Clarity

Tesla Roadster

Great Wall Smart EV

FIAT Phylla

Chevrolet Volt

BYD F3DM

- Extended-range electric vehicles (EREVs) like the Chevrolet Volt, BYD F3DM, and Daimler Blue Zero E-Cell, which will have battery-only ranges of around 40 miles, satisfying the daily driving needs of three out of four urban dwellers.
- Fuel cell electric vehicles (FCEVs) like the Honda FCX Clarity and Chevrolet Equinox Fuel Cell, which can be driven more than 200 miles between fill-ups of hydrogen.

Battery-electric vehicles produce zero emissions and promise to be the most affordable solution for limited-range urban applications. Battery-powered vehicles become even more compelling when one considers that lithium-ion battery packs can be recharged in less than three hours using a 240-volt electrical outlet, or overnight in about eight hours using a standard 110-volt outlet. The cost of that recharge is very low, only about 2 cents per mile for

	Battery-electric vehicle	Extended-range electric vehicle	Fuel-cell-electric vehicle
Vehicle size	≤ Small	≤ Compact	≤ Family
Refueling time	Hours	Hours (battery charging)	Minutes
Range (miles)	100+	40 (battery)/ 300 + (overall)	300–400
Performance	Excellent	Excellent	Excellent
Vehicle emissions	Zero	Zero for 40 miles daily	Zero
Energy source	Diverse/ petroleum free	Diverse/petroleum only with range extender	Diverse/ petroleum free
Refueling infrastructure	Already available at home	Already available at home and stations	Must be deployed

Figure 2.4
Characteristics of electric-drive systems.

the electricity, or about 80 cents per day—which is far less than the cup of coffee that many of us buy on the way to work. This is also one-third to one-sixth the cost of driving a comparable vehicle fueled by gasoline, depending on whether the price at the pump is $2 or $4 per gallon.

An extended-range electric vehicle, such as the four-passenger Chevrolet Volt sedan, has an engine/generator that kicks in and allows continued electric-drive operation on those days when more than 40 miles are driven. This range-extender can be fueled by gasoline or biofuels and increases the range by hundreds of miles, addressing the "range anxiety" associated with pure battery-powered vehicles but adding complexity and cost to the vehicle.

Only the hydrogen fuel cell option promises to combine the range and refueling-time convenience of conventional family-sized vehicles with the energy and environmental benefits of pure battery-powered vehicles. Moreover, the hydrogen

infrastructure complements the electric grid from an energy-diversity perspective because domestically supplied reformed natural gas and biomass are excellent sources of hydrogen, and hydrogen, generated from electrolysis of water, is an excellent way to store electricity produced from renewable sources like the wind and sun.

General Motors, Daimler, Honda, Toyota, Ford, and others have been working over the past decade to develop fuel cells as a viable automotive propulsion option. Collectively, these manufacturers have now achieved an impressive list of benchmarks related to fuel cell vehicle performance, safety, range, speed, cold-start capability, durability, and functionality.

Figure 2.4 provides a comparison of the three types of electric-drive vehicle, showing how they stack up on a variety of metrics and illustrating that each has a unique "value proposition," meriting their continued development. Battery-electric, extended-range electric, and hydrogen fuel cell electric vehicles are complementary; all have promising futures, and all offer a range of value propositions for automobile customers while diversifying energy use and significantly enhancing vehicle efficiency and emissions on a total "well-to-wheels" basis.

The role that electric-drive vehicles can play in the portfolio of vehicle segments is illustrated in figure 2.5. For family-sized vehicles, there is great potential to realize long-range zero-emissions driving with rapid refueling if the hydrogen fuel cell is chosen as the power source. At the other extreme, batteries can provide the onboard electricity for small urban vehicles with limited range. Extended-range electric vehicles fill the gap between these two approaches and could be the best choice for the family sedan.

Another feature of electric-drive vehicles is that they can be driven either by central motors—which would suffice for most vehicles—or by motors in the wheels at each corner. The most promising application for wheel motors is likely to be in urban vehicles because their need to be compact places a premium on the space liberated by moving the motors outboard into the wheels.

Figure 2.5
Some applications of electric-drive vehicles.

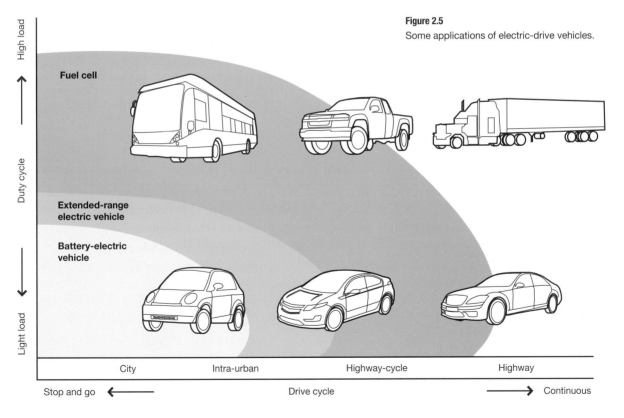

High load

Duty cycle

Light load

Fuel cell

Extended-range electric vehicle

Battery-electric vehicle

City   Intra-urban   Highway-cycle   Highway

Stop and go ⟵   Drive cycle   ⟶ Continuous

Electric-drive also creates the foundation to electrify other vehicle systems, such as so-called chassis-by-wire systems. These allow the chassis to follow the driver's control input rigorously. With actuators for steering, brakes, and suspension, a co-ordinated chassis controls approach permits an integrated chassis response to the driver's command. The Chevrolet Sequel (figure 2.6) was the first roadworthy vehicle to demonstrate full electrification of vehicle systems (electric-drive, including rear-wheel motors, and by-wire braking and steering).

By-wire systems also provide design, manufacturing, and performance benefits. For example, brake-by-wire creates space in the vehicle by eliminating the brake master cylinder, booster, hydraulic lines, and park brake. From a manufacturing perspective, brake-by-wire will simplify the assembly of corner modules and reduce assembly time for the brake subsystem. And it enables performance improvements. For example, when the brake pedal is pressed, the response is like switching on a light, and the vehicle comes to a stop sooner. This can help to reduce stopping distance and avoid many collisions. Other benefits include the ability to tune the brake-pedal feel and to integrate with throttle, suspension, and steer-by-wire systems to enhance stability and active safety (collision avoidance).

Steer-by-wire offers similar types of benefits to brake-by-wire (design, manufacturing, and performance). For example, by eliminating the intermediate shaft, it offers greater front-compartment packaging flexibility. Because there is no intermediate shaft, assembly is simplified between the body and chassis. Stability and handling are improved because the wheels can be steered independently of each other and of the steering wheel. Steer-by-wire is more adaptable than conventional steering, so that in a parking lot it can be very sensitive whereas on the highway it will be more stable (i.e., speed-sensitive). In addition, noise and vibrations are reduced owing to the isolation of the steering wheel by intelligent filtering of road feedback.

By-wire controls can be upgraded with software for consumers who desire a higher-performance chassis or a newly styled body. For example, software changes can allow the chassis settings (e.g., brake pedal feel, steering effort and feedback, and suspension ride) to be automatically adjusted to the different bodies that dock onto the same chassis hardware or according to the driver's preferences. Preferred settings may be downloaded from a menu of allowable calibrations online and accessible from one's mobile phone. Chassis personalization will make electrified vehicles popular not just with environmentalists but also with performance enthusiasts. Sounds, for example, can be electronically generated by the electric-drive system to match the person's mood while accelerating, like ringtones for mobile phones.

Until now, it has been difficult to commercialize chassis-by-wire systems because the conventional DNA's 12-volt starter battery has insufficient juice to power systems that activate brake and steering systems; but with the new DNA's higher-voltage electrical systems it will be possible to realize the many benefits of by-wire chassis.

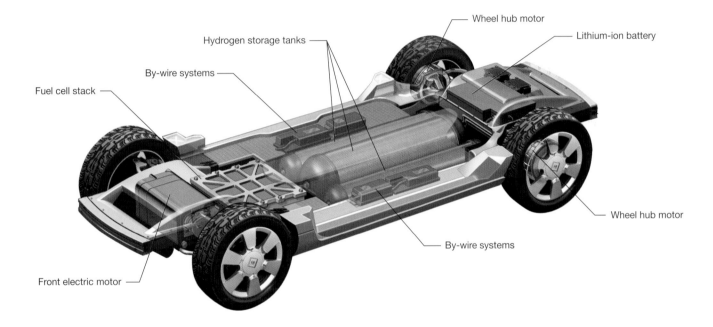

Wheel hub motor

Lithium-ion battery

Hydrogen storage tanks

By-wire systems

Fuel cell stack

Wheel hub motor

By-wire systems

Front electric motor

Figure 2.6
The skateboard of the fully electrified, drive-by-wire
Chevrolet Sequel.

The New DNA (Part 2): Connectivity

The other enabler of the new automotive DNA is connectivity among vehicles. Think of future vehicles as souped-up driving platforms that are integrated with wirelessly networked computers on wheels. These vehicles will be accurately located using GPS technology. They will have the capability to sense objects all around them. They will use wireless systems to communicate with other vehicles and with the roadside infrastructure. Eventually, they will even be able to drive themselves and automatically avoid crashes.

There is a close analogy between these sorts of networked personal mobility devices and networked personal computers. Since their introduction in the early 1980s, personal computers have improved along all dimensions, but it was their interconnection through an infrastructure (the Internet) that most dramatically increased their functionality and appeal.

Many of the technologies necessary for comprehensive vehicle connectivity already exist in commercial form. Existing vehicle telematics systems, such as OnStar, combine GPS, digital maps, and wireless communications to provide features such as hands-free phones, real-time turn-by-turn navigation and traffic information, remote diagnostics, automatic crash notification to emergency response authorities even if the driver is incapacitated, and stolen vehicle slowdown. In parallel with the development of vehicle connectivity, sensor-based features are also being introduced into automobiles. These include blind-spot detection systems that can sense objects not normally visible to the driver and warn the driver of their presence, and lane-keeping systems that maintain the vehicle's lateral position so that it does not unintentionally drift into another lane.

What can be achieved when GPS, digital maps, and sensing are combined was demonstrated in November 2007, when eleven self-driving, or autonomous, vehicles successfully competed in the Defense Advanced Research Projects Agency (DARPA) Urban Challenge, a prize competition for driverless vehicles. The winner is shown in figure 2.7. Since one of the competition's rules was to forbid communications technology (as this could be jammed in a military confrontation), the vehicles had to rely solely on GPS and sensing technology for environmental awareness and collision avoidance. Even with this limitation, four teams still completed the sixty-mile race by driving autonomously in a mock urban environment that also featured conventional vehicles being driven by people.

The DARPA Urban Challenge Vehicles were all large and expensive. However, there are numerous opportunities to reduce the bulk and cost of their electronic systems. Eventually, autonomous driving will become affordable in small, mass-produced vehicles.

An automobile can know the location and speed of an approaching vehicle by sensing it (with radar, for example). An alternative solution is to communicate wirelessly and exchange information with the approaching vehicle. Vehicle-to-vehicle (V2V) communications is essentially a new GPS- and wireless-based all-around object-detection sensor that gives the vehicle a "sixth sense." With V2V, each vehicle broadcasts its position and velocity to its "neighborhood" and continuously monitors the status of surrounding vehicles within a quarter-mile radius. This provides the vehicle with an "extra set of eyes" so that it can see its local environment and can detect road situations around a corner or even ten vehicles ahead. Prototype V2V communications systems have been demonstrated that support automated safety features like lane-change alert, blind-spot detection, sudden stopping, forward collision warning with automatic braking, and intersection collision warning. These have demonstrated a very robust performance that is immune to extreme weather conditions.

Both sensing and communications have advantages and disadvantages. Sensing adds hardware mass and cost to the vehicle, and the vehicle may not be able to sense an object in poor visibility conditions, or if there is a vehicle approaching around a corner that is not in its line of sight. Relying on communications, however, requires that there be another vehicle or object nearby and that it can respond back. Again, the two approaches are complementary, and they may need to be combined to provide the most cost-effective and robust solution for a wide range of real-world conditions. Figure 2.8 contrasts the existing approach to sensor-based collision avoidance with the future version that combines wireless communications and sensing to provide a lower-cost, more functional safety system.

In the future, more sophisticated vehicle-to-vehicle and vehicle-to-infrastructure communications will enable automated, cooperative driving. Vehicles

Figure 2.7
The Chevrolet Tahoe "Boss" autonomous vehicle, winner of the 2007 DARPA Urban Challenge. The vehicle was fielded by a team led by Carnegie Mellon University.

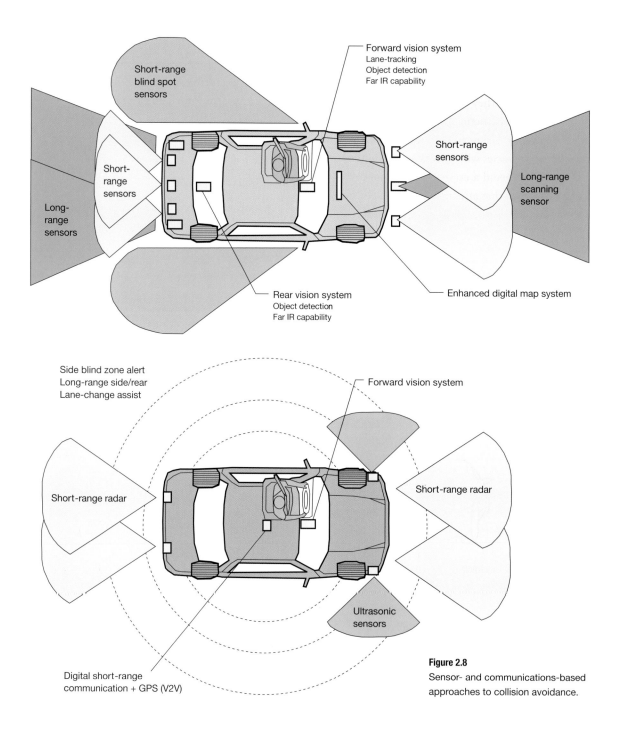

Short-range blind spot sensors

Forward vision system
Lane-tracking
Object detection
Far IR capability

Short-range sensors

Short-range sensors

Long-range scanning sensor

Long-range sensors

Rear vision system
Object detection
Far IR capability

Enhanced digital map system

Side blind zone alert
Long-range side/rear
Lane-change assist

Forward vision system

Short-range radar

Short-range radar

Ultrasonic sensors

Digital short-range communication + GPS (V2V)

Figure 2.8
Sensor- and communications-based approaches to collision avoidance.

will be accurately located using GPS technology and will be able to sense objects all around the vehicle, through some combination of sensing and wireless communications. They will also communicate with the roadside infrastructure and will even be able to drive themselves. Using wireless transponders, these autonomous vehicles will sense what's around them and will either avoid a crash or decelerate to a low enough speed so that any impact will not be harmful to pedestrians, cyclists, and vehicle occupants. The transponder (one has already been prototyped by General Motors) continues to shrink in size to a point where it can fit inside a pocket. At the size of a cell phone, transponders can be carried by pedestrians or cyclists, to increase the safety of these vulnerable road users. This capability can eliminate most of the vehicle collisions that cause injury or property damage (figure 2.9).

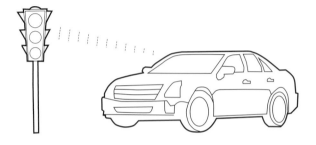

Vehicle to roadside infrastructure

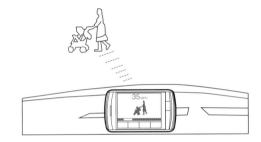

Vehicle to pedestrian (with transponder)

Vehicle to vehicle

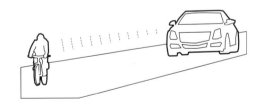

Vehicle to cyclist (with transponder)

Figure 2.9
V2X includes vehicles communicating with each other (V2V), the roadside infrastructure (V2I), and pedestrians and cyclists (V2P).

Synergies between Electrification and Connectivity

Electrified and connected vehicles separately offer compelling sustainability benefits; when they are combined the benefits are even greater. Electrification provides precise and responsive actuation, whereas connectivity provides situational awareness. Powerful information processing underlies both, and taken as a whole, this new DNA offers a far superior sensing-processing-actuating system than the current automobile's DNA. This enables efficiencies at the level of vehicle design, and could also make future vehicle traffic behave much more like the highly efficient swarms seen in nature.

Design freedoms resulting from connectivity-enabled zero-crash capability provide many opportunities to take cost and mass out of the vehicles, since it will not be necessary to meet the same high-speed crash requirements as conventional automobiles. Structures, already lightened by only needing to support a modest propulsion system, can be put on an even stricter diet if they do not have to provide all the traditional crash-resistance features. It may be possible to realize vehicle mass savings in excess of 20 percent if safety-related content can be avoided, which could translate into a fuel economy improvement of more than 10 percent. Although this mass reduction would also apply to conventional automobiles, the resulting lower energy consumption is more critical for enabling the new DNA, as it reduces the cost and mass of battery, hydrogen storage, and electric motor systems, which in turn further reduces overall vehicle mass. Lower

vehicle mass also yields additional safety benefits through improved responsiveness and lower inertial impact between similar sized vehicles. Moreover, the packaging freedom associated with both electric-drive (no engine compartment) and crash elimination (minimal front-end crash structure) can help to make vehicles significantly shorter (and again, therefore lighter), which is particularly useful where parking space is limited and expensive.

The synergy emerges in other ways as well. For example, one of the challenges associated with battery-electric and hydrogen fuel cell electric vehicles is the initially limited availability of nearby charging and hydrogen-refueling stations. Connectivity can provide drivers with real-time information on locally available hydrogen-refueling and battery-charging stations and can, therefore, play a significant role in making drivers more comfortable with purchasing and using electric-drive vehicles.

We already have clear evidence that connectivity can be used to smooth out traffic flow. EZ-Pass toll roads reduce the need for stopping and starting. The coordinated timing of traffic lights is another example of the benefits of an intelligent infrastructure, as it has repeatedly been shown to reduce the frequency of stops, raise average speeds, and increase fuel economy of vehicles by more than 10 percent.

Greater awareness of the external environment when driving, through wireless communications, has huge potential to improve traffic flow, reduce the number of crashes, and reduce air pollution and energy consumption. These are interrelated;

a reduction in crashes should increase traffic flow, which will reduce idling losses and raise energy efficiency. These efficiency improvements, enabled by connectivity and intelligence, can further reduce the size, weight, and cost of hydrogen storage, fuel cells, and batteries.

Vehicles can also platoon together, using wireless communications to maintain a constant, close separation. Tests performed in the 1990s at the University of California, Berkeley, concluded that by reducing drag coefficients (as much as 25 percent) and transient operation, energy efficiency could be increased 10–20 percent with platooning separations of under 10 meters (about two vehicle lengths). Vehicles could drive together as a system, but the occupants of each vehicle would still be free to separate whenever they wished.

Platooning also allows a useful new form of vehicle modularization. Compared to a conventional small family car, for example, two smaller two-passenger vehicles would provide a family with far greater freedom of movement. It is also easier to find parking for two smaller vehicles than for one bigger vehicle of the same overall size.

Furthermore, connectivity allows greater knowledge of the journey ahead (whether a hill, a traffic light, or arriving home), and this can help improve fuel economy even for conventional vehicles. In urban environments, significant amounts of fuel can be consumed in the search for parking, and knowing the location of available parking spots can help to reduce unnecessary driving and increase the vehicle's useful range. Having knowledge of traffic conditions

in real time can also alert the driver to congestion along the normal route and give the driver a choice of finding an alternative, faster route that avoids the traffic buildup; the earlier this warning can be given, the greater the potential for fuel economy improvement. With an electric-drive vehicle, the opportunity for improving the already higher fuel economy is even greater, because there is more flexibility to shuttle energy between the battery and the engine/generator or fuel cell. For example, if the vehicle has a navigation system and knows it is returning home, or climbing a hill with a subsequent descent, then the vehicle's propulsion control system may permit greater depletion of the battery than it would otherwise allow since it knows that the battery will be recharged soon, either at home or when going downhill.

Connectivity can help raise fuel economy not only by improving traffic flow but also by providing feedback on how to drive more efficiently by comparing the current driving performance with the previous month's driving or with other people's driving on similar routes. Both approaches can raise efficiency by at least 10 percent.

In summary, electrification and connectivity will provide enhanced personal mobility in terms of faster trips and reduced variation in trip time, easier maneuverability and parking, and a more personalized ride experience. By facilitating smoother flowing traffic with less congestion, these technologies will make travel times more predictable and average speeds and energy efficiency higher, making them more conducive to electric-drive, batteries,

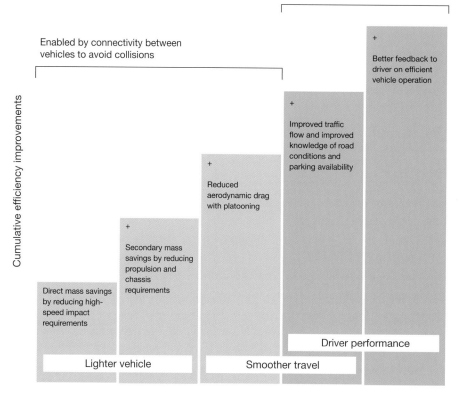

Figure 2.10
Energy efficiency benefits, arising from the synergy
between electrification and connectivity.

and hydrogen storage. The design freedom enabled by electric-drive and cars that do not crash will, in turn, make it possible to design smaller, lighter, more efficient cars, which will also be more affordable as a result. Figure 2.10 provides a qualitative summary of the synergistic efficiency benefits that can be achieved when electrification and connectivity are combined.

The Emerging Personal Mobility Revolution

As with the digital revolution, the personal mobility revolution will be realized by the convergence of a host of emerging technologies, not by any single technological innovation. And just as the Internet does not give one manufacturer an advantage over another but makes all computers more functional, the same will be true with this new smarter mobility network. The new DNA of the automobile, integrated within this system, will make driving more practical and enjoyable. It will make the future automobile a desirable item that provides style, status, and an emotional response.

Importantly, the new DNA transforms the automobile from simply a means to get from point A to point B to a node in a global network that integrates flows of vehicles, information, and power. The automobile's shift from a stand-alone, largely mechanical device to one that is electrical and connected will be every bit as momentous as the transition from horses to horsepower that occurred a hundred years ago.

The Opportunity for a Design Renaissance

To succeed on a large scale, future vehicles must have the look and feel of a new and desirable kind of product. Nobody thinks of an iPod as a shrunken home stereo system, and nobody should be left with the impression that an intelligent electric-drive vehicle is a dull but worthy "econobox." To achieve efficiencies at the scale needed, automobiles need a new type of architecture that builds on the DNA we have described to shift as much mass as possible from the moving vehicle to the fixed infrastructure, takes advantage of opportunities to miniaturize and lighten hardware, exploits weightless software wherever possible, and combines systems to eliminate redundancy. As we have seen, key enablers of this new architecture will be electrification and connectivity. All this yields many opportunities to introduce new materials and details. Automobiles with this new architecture can contain much less material and have a great deal less exterior surface area. The use of light, high-performance, but more expensive materials therefore becomes feasible. Painted sheet metal is no longer a given.

A new architecture, such as the one we propose, creates opportunities for innovative designers. When products are highly evolved, regulated, and marketed under conditions of intense price competition—as with today's standard automobiles—the opportunities for major design innovation are very limited. Designing them beautifully requires great subtlety and skill, but it often reduces to variation of a well-defined, well-established type. But whenever a radically new architecture emerges, there are opportunities to explore hitherto unimagined technical solutions and forms. With the reinvention of automobile DNA, we have the conditions for an automobile design renaissance.

The Electric Skateboard Concept

A crucial feature of the new architecture we propose is the electric "skateboard," derived from eliminating the traditional engine and drive train and packaging batteries or hydrogen fuel cells in innovative ways.

The horseless carriage replaced the horse with fuel tank, engine, and mechanical drive train, and replaced reins with mechanical and hydraulic controls. Today's hybrids and battery-electric vehicles keep essentially the same vehicle architecture, but introduce into it batteries and electric motors. Although this keeps development and manufacturing costs down, the end result is that vehicles look pretty much the same, differing primarily only in the amount of fuel they consume and the amount of emissions leaving their tailpipe. Even purpose-built electric-drive vehicles, like General Motors' EV1, have embraced the same basic DNA as a conventional automobile by retaining the front (engine) compartment, instrument (dash) panel, foot pedals and steering wheel, fixed seating, hydraulic brakes and steering, and so on.

In the near future, when batteries and hydrogen fuel cells have displaced the gasoline engine, automakers will design vehicles around these electrochemical devices (just as vehicles today have been designed around an engine, itself a throwback to the days of the horse and carriage). What will this mean in practice?

Electric drive opens up more freedom for designers because, compared with combustion engines, it provides much greater flexibility in designing the shape of batteries and fuel cells. For example, they can be shaped like cylinders or boxes (a "suitcase" or a long rectangular chain, perhaps), and this flexibility allows engineers and designers more options for locating them in the best place to improve the vehicle's styling and all aspects of the vehicle's performance. If these electrochemical devices are mounted under the floor, they can provide a flat "skateboard-like" foundation that creates a clean sheet for exterior styling, since any shape above the plane of the flat rolling chassis might become possible, as indicated in figure 2.11. Eliminating the engine and the engine compartment makes it possible to imagine new shapes and proportions. By placing the propulsion system completely inside the chassis frame we can let the driver experience an SUV-like command-of-the-road seating position while enjoying the surefootedness normally found in vehicles having a much lower center of gravity.

General Motors' "AUTOnomy" concept embodied the electrification of the automobile and the skateboard architecture, but retained the conventional four-wheel layout. The skateboard architecture can easily be extended to other wheel configurations, such as a three-wheeler or the two-wheel tandem (two wheels in-line) layout seen in bicycles and scooters. It can also enable new features, such as folding mechanisms and balancing machines, as shown in figure 2.12.

Skateboard automobiles with by-wire control can do away with the conventional mechanical steering column and hydraulic brake systems that force today's mechanically driven vehicles to have a fixed location for the steering wheel and foot pedals (or human–machine interface, to use technical jargon) and an interior design that has barely changed in the last hundred years. In contrast, by-wire electrical systems can be linked together very easily, as anyone moving cables around behind a computer or TV has experienced. This can mean far more opportunity to blend the driver's controls with the vehicle's design personality. In other words, if a vehicle evokes jet aircraft imagery, then we might want it to have a way to accelerate by pushing the "steering wheel" forward and to brake by pulling it inward. Generalizing, a sports-car driver might want a different way to steer, brake, and accelerate than the owner of a conservative, luxury vehicle. Until now this has not been easy to do, but with by-wire systems this will change.

Another plus for the skateboard, when combined with by-wire control, is passenger comfort. In most of today's vehicles, the rear-seat passengers have modest amounts of legroom, while the driver is constrained because the right foot needs to be "tethered" to the pedals, the steering wheel is in front of the stomach, and the head is close to the roof. By-wire systems can eliminate pedals and allow pure hand controls, making it possible to stretch one's

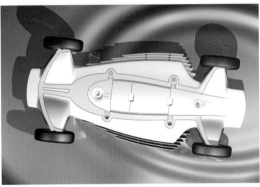

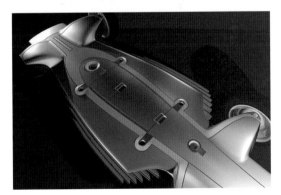

Figure 2.11
The AUTOnomy skateboard and body.

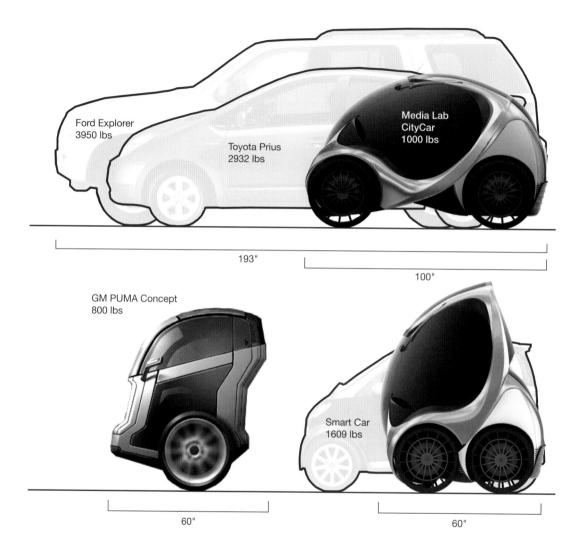

Ford Explorer
3950 lbs

Toyota Prius
2932 lbs

Media Lab
CityCar
1000 lbs

193"

100"

GM PUMA Concept
800 lbs

Smart Car
1609 lbs

60"

60"

Figure 2.12
Folding and balancing to reduce vehicle footprint.

legs and to move them freely. The elimination of the engine compartment can increase the usable length of the interior and enable more legroom between the rows of seats.

Since in these configurations the driver does not need to look over an instrument panel or engine compartment, and because the lower leg does not need to be nearly vertical (because there is no brake pedal), it may be possible to recline the front seat more than we can do today. Reclined seats would help to lower the vehicle's height and create more aerodynamic designs and sharper styling. All of these ideas, together with a steering pod that could slide from left-hand drive to right-hand drive, were executed on the Hy-wire, introduced in September 2002 at the Paris Mondial (figure 2.13). This vehicle, the first drivable vehicle to combine an electric-drive system and by-wire chassis controls in a skateboard chassis, vividly demonstrated the potential for rethinking an automobile's interior space.

Although an electric skateboard provides tremendous flexibility in the vehicle's body styling, the by-wire technology may offer even more. By reducing the number of links between the interior and the rolling chassis, it allows automobile bodies to be swapped on top of a rolling chassis—which means easier upgrades or reconfigurations.

When skateboard-based vehicles do not have to be designed to withstand high-speed crash events, there will also be many opportunities to rethink structure, surfaces, and glazing. Their bodies could be constructed from metal frames or be composite exoskeletons combined with lightweight polycarbonate

Figure 2.13
The interior layout of the Hy-wire showing control from either the left or the right seat.

panels, for example, instead of sheet metal and glass. These panels might be tinted or fritted to reduce light and heat transmission, and they might actively adapt to changing exterior conditions by means of embedded sensors and LEDs, photochromic material that darkens in response to bright sunlight, and electrochromic material with digitally controllable transparency. This results in less demand on cabin heating and cooling systems, and so enables further mass reduction.

Within the cabin, safety content such as airbags, padded interior surfaces, and seat attachments to the floor could be simplified or eliminated, yielding further weight savings. Seats could be much lighter weight, and the instrument panel could be reduced in size or even eliminated. Air conditioning systems could be significantly smaller and more efficient because they would not have to be designed to cool down such a large, thermally massive interior as in a conventional automobile.

Novel Ways to Get In and Out

The basic architecture of an automobile determines how people get in and out of it. You can get into a standard automobile from its two sides. You can't get in from the front because the engine is there. Even in rear-engine cars, you normally cannot enter from the front because the steering wheel and the dashboard get in the way—some rare exceptions, like the Isetta, aside. Another possibility, rarely seen in everyday passenger cars for obvious reasons, is to clamber in from the top, as into a fighter plane

or racing car cockpit. The standard arrangement allows angle parking nose-in to a wall, but requires side- and rear-access clearance. It also allows parallel parking against a sidewalk, but with the disadvantage that the driver-side doors open into the traffic stream—so that exiting passengers must proceed with caution in the face of oncoming vehicle and cyclist traffic.

Since they don't have central engines or motors to get in the way, and can replace traditional steering wheels and dashboards with less obstructive electronic devices, future automobiles could allow entry and exit (ingress and egress) from all four sides. However, providing entry and exit from all four sides would increase complexity, cost, and clearance requirements.

A more attractive option is to combine front entry and exit for passengers with rear entry and exit for baggage and emergencies, as shown in figure 2.14. For street parking, in combination with decreased length and increased maneuverability, this allows cars to park nose-in to the curb. Side clearance can be minimal. This also allows more than three times as many cars to be parked along a given length of curb than with normal parallel parking, without increasing the width of the parking area. It also enhances convenience and safety, since the driver can step out directly onto the sidewalk. In domestic garages and in parking structures, vehicles can either park nose-in to pedestrian walkways, thus separating pedestrian and vehicle zones, or can rotate and park rear-in to minimize parking space requirements.

Figure 2.14
Entry and exit through the front.

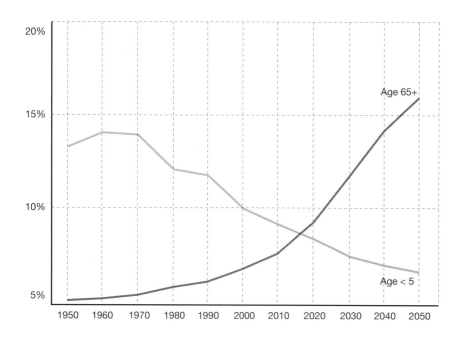

Figure 2.15
The aging of the world's population.

Entering or leaving the vehicle through the front should make for an easier experience than getting into a sedan. This will be particularly important for the aged, which means more and more of us, as demographically the population is aging. By 2025 more than a third of Japanese citizens will be over 60 years of age, and there will be 290 million people in China aged 60+—about the same as the total U.S. population today. Within the next decade, the number of people in the world aged 65 or older is expected to exceed the number of people under the age of 5 for the first time in history, as shown in figure 2.15. These same aging drivers will also value the independence that autonomous vehicle operation can provide them, reducing the need to rely on others for mobility.

There is no longer any reason to let the exigencies of drive-train engineering dictate the way we get in and out of cars. We can, instead, develop ingress and egress systems that make ergonomic and urban design sense.

Summary: Reinventing the Automobile

In its first hundred years, the automobile did more than revolutionize transportation; it also shaped our society. In its second hundred years, even as we shape our vehicles to society's requirements, they will continue to drive the future. Through electrification and connectivity, we will have the opportunity to create "city" cars for urban dwellers, autonomous vehicles for the young and old, and clean, efficient vehicles that greatly reduce the energy, environmental, safety, and congestion problems associated with the automobile. The new DNA will enable the transformation of our vehicles into interconnected nodes in transportation, power, information, and social networks, providing new business options and opportunities that will spur sustainable economic growth around the globe.

As with the digital revolution, the personal mobility revolution will be realized by the convergence of a host of emerging technologies, not by any single technology solution. And just as the Internet does not give one PC manufacturer an advantage over another but makes all computers more functional, the same will be true with our future mobility systems. The new DNA of the automobile, integrated within this system, will make driving more practical, enjoyable, and rewarding for everyone.

Vehicles based on the new DNA will have outstanding market potential because they offer the opportunity to drive innovative automobiles that are smart, responsive, fun to take out on the road, and exciting to own. They will be simpler and have fewer parts, making them more affordable than today's cars and trucks. And they will be safer, cleaner, and more energy efficient.

The Mobility Internet

For over half a century, learning to drive an automobile has been seen as a rite of passage. It has come to symbolize a newfound freedom to go where we want, when we want, and with whom we want. In the last few years, another consumer product has achieved the same status: the mobile phone. Preteens desire it for the same reason they later desire an automobile: it provides them with a chance to connect with their friends wherever they are and at any time.

Although the convergence of communications technologies (such as the merging of mobile phones, personal digital assistants, GPS navigation systems, MP3 players, and so on) is well advanced and is providing people with unprecedented levels of connectivity, one huge gap remains. This makes itself evident when we are driving. Ironically, when the ubiquitous connectivity provided by mobile phones is added to the freedom of mobility provided by the automobile we get *less* than the sum, because using mobile phones distracts drivers and increases the likelihood of accidents. However, government restrictions on mobile phone usage (such as no texting and hands-free calling laws) tend to be unpopular because they limit our freedom to do what we want when we want. Extension of the Internet to encompass intelligent automobiles will rectify this (and a good many other current problems and limitations) by allowing vehicles and their occupants to be connected safely and efficiently with each other, with the roadside infrastructure, and with central servers.

Let's call the result of this the Mobility Internet. Like the Internet we know today, it will move huge amounts of data around in real time over vast areas, but it will also coordinate the movements of people, vehicles, and goods. Vehicles will become network nodes on wheels; they will acquire, process, utilize, and communicate information that supports their individual functions and those of the mobility system as a whole; and they will be routed efficiently from place to place much like packets of data in the Internet.

There are many precursors to the Mobility Internet and precedents for it. Ships, airplanes, and automobiles were among the first beneficiaries of wireless technology—both broadcast and two-way. Mobile phones quickly found a home in automobiles. GPS navigation systems depend on wireless links to satellites. Electronic toll systems rely on transponders and readers. Systems such as GM's OnStar provide security, emergency response, diagnostic, and other services via wireless connections to service centers. But these capabilities are largely add-ons to automobiles that were designed independently of them and could function quite adequately without them. Like the Internet that has become so familiar to us, the Mobility Internet will increasingly become the unified delivery mechanism for data streams and services that had previously been separate. Vehicle designs will assume and depend on onboard intelligence combined with sophisticated mobile connectivity. And these capabilities will enable the integration of automobiles into urban-scale networked computing and control (NCC) systems.

These systems will efficiently manage traffic flow, safety, road space, parking space, vehicle fleets, and electric supply.

There are two distinct but interrelated aspects to the Mobility Internet: networked computing and control for vehicles and social networking for their occupants. We will consider these in turn.

Networked Computing and Control

There are many technical challenges in implementing networked computing and control for urban automobiles and personal mobility systems, but effective technological solutions are emerging and converging.

The first and most obvious challenge is to provide sufficiently fast, reliable, two-way connectivity to large numbers of geographically dispersed, moving automobiles. One possibility is to use existing, subscriber-based cellular networks, but this becomes expensive at a large scale. A more recently emerging possibility is to create citywide mesh networks in which moving vehicles can connect to one another and opportunistically take advantage of nearby wireless access points in buildings and fixed infrastructure.[1] This sort of connection is fleeting and intermittent, but it is inexpensive, it can provide broadband speeds, and it can be managed through use of appropriate protocols and software.

The second major challenge is that of scalability. It is technically straightforward to connect a few vehicles and collect data from them, but modern urban environments may contain millions of vehicles, smart mobile phones, and roadside sensors. All of these can potentially be used for harvesting data about vehicle and pedestrian movements, local road and weather conditions, and so on. The result is a huge stream of data—billions of time-stamped, location-tagged sensor readings—which must be transferred to servers, inserted into databases, queried and processed, and acted on in real time. Standard database techniques are not well adapted to this task.

The third challenge is that of massively distributed computation and control. Vehicles receive, process, and act on data from their sensors, from other vehicles, and from remote servers. Simultaneously, servers receive data from dispersed vehicles, process it, and send signals back to vehicles. All this must add up to an effective, efficient, real-time control system for personal urban mobility systems. As we shall see later (particularly in chapter 8), such systems can achieve great efficiencies by smoothing out traffic flows and by regulating supply and demand for road space, parking space, vehicles, and electricity.

Finally, there is the challenge of accomplishing all this while preserving locational privacy. Drivers will want to benefit from networked computing and control, but they will not want to surrender their privacy in order to do so. Fortunately, this should not be necessary. If networked computing and control systems for personal urban mobility incorporate appropriate cryptographic protocols, then they should be able to perform their functions without violating locational privacy.

Trip Times and Congestion

The need for networked computing and control arises largely because vehicles do not operate independently of one another. (The lone vehicle on the open road is an exceptional limit case.) Along streets and roads, they combine to form traffic streams. Furthermore, these streams cross, subdivide, and combine in complex ways. Under ideal conditions, traffic streams flow smoothly. But when traffic volumes approach roadway capacity limits, and when interruptions occur, their behavior becomes notoriously chaotic and unstable, with stops and starts, slow periods, and traffic jams.

The problem of maintaining optimal separation and speed is difficult to solve when it is left primarily to human drivers, since humans have limited information-processing capacity, are easily distracted, and react in psychologically complex and sometimes irrational ways. The problem is exacerbated by the demands of merging, changing lanes, negotiating intersections, and responding to unexpected obstructions.

Extensive studies of traffic flow have demonstrated that even small variations and interruptions, especially when multiplied in heavy traffic, can propagate shock waves back through traffic streams for miles. And when traffic jams begin to form, they tend to get worse as too many cars encounter too little road throughput. This chaos wastes time, space, and energy. And it is dangerous, resulting in traffic accidents, injuries, and deaths.

The time burden imposed by congestion is caused by both the extra journey time it adds and

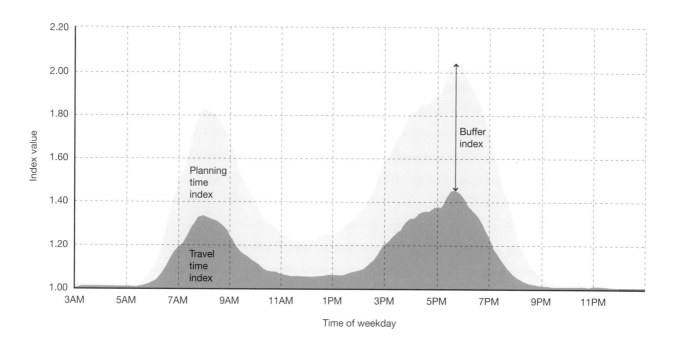

Figure 3.1
Travel "buffer" time budgets.

the unpredictability of the trip length it imposes, which means more "buffer" time is often allotted to a journey than is typically needed. As figure 3.1 shows, when traveling in rush hour people typically budget twice as long for a trip as it would reasonably be expected to take under noncongested conditions, whereas statistically it may only require 50 percent more time to travel under those conditions.

Choreographing the Highway

Can networked computing and control address the causes of congestion so that traffic can be managed more effectively and accidents avoided or even eliminated? Can it be used to optimize traffic streams to reduce this form of variation and waste? The most effective approach is to make use of electronics and software to emulate swarms of ants, flocks of sheep, even pedestrians at a busy intersection.

The Shibuya crossing in Tokyo may be the busiest pedestrian intersection in the world, but it is rare for people to collide with each other when crossing, even if the crowd is so large that it covers the entire road surface. In three dimensions one can also marvel at locusts that manage to swarm without crashing into each other. What these examples show is that in nature the sensing, processing, and actuating capabilities exist to prevent the types of collisions and accidents that routinely occur with vehicles.

Future vehicles will approach what nature accomplishes, and in analogous ways, through use of greater sensing power, increased communications and processing bandwidth, and more precise actua-

tion. Streets and roads will be electronically choreographed through use of increasingly precise GPS and other location technologies, onboard sensing, and wireless communications.

Dedicated Smart Vehicle Lanes

But can we achieve this level of coordination purely by making vehicles smarter? What happens when that classic 1950s Corvette cuts in between these future smart vehicles? What happens when there is black ice ahead? Or when a child runs out of nowhere onto the street?

These scenarios and many others make a case for separating types of traffic, at least in the short term (figure 3.2). Governments around the world are increasingly recognizing this and creating dedicated lanes, sometimes with physical buffers which provide cyclists with additional protection from automobiles. Dedicated lanes can also be provided for automobiles (HOV—High Occupancy Vehicle, or HOT— High Occupancy Toll) or for other forms of mobility (BRT—Bus Rapid Transit, for example). Abu Dhabi's Masdar city is going even further and incorporating a dedicated layer and grid network for Personal Rapid Transit. However, the advantages of separation must be weighed against the disadvantages of requiring more road space—which will not always be available—and of increasing the complexity of intersections.

Where dedicated lanes or zones are used, smart vehicles could electronically identify and authenticate themselves prior to accessing them. They could

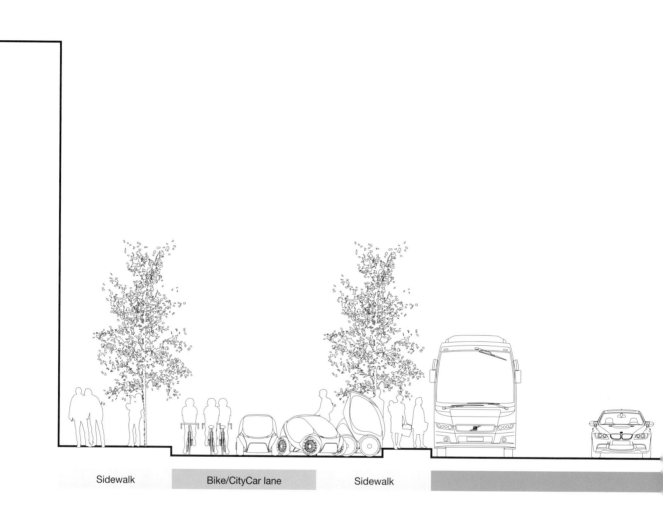

Sidewalk   Bike/CityCar lane   Sidewalk

Figure 3.2
Urban street, showing separation for pedestrians,
bicycles and light vehicles, cars, and buses.

| Car lane | Sidewalk | Bike/CityCar lane | Sidewalk |

then be protected from other road users by physical barriers. These lanes might not only provide more controlled environments, but also specialized electronic infrastructure, and even roadway recharging strips as discussed in chapter 6. Over time, as the percentage of smart vehicles on the roads grew, the numbers of lanes dedicated to them might be increased while the numbers dedicated to conventional vehicles were correspondingly reduced.

Providing Warnings

If sensors and communications capability can be incorporated at key points alongside the road, such as at intersections or at troublesome geographic spots where ice is likely to form easily, then many accidents can be eliminated. U.S. data indicate, for example, that the economy suffers $3 billion of damage each year because drivers accidentally run through a red light (because of some kind of distraction) and cause an accident. This type of accident can be avoided if the intersection, as shown in figure 3.3, sends a signal to the vehicle and provides a visible warning to the driver that the light is red (or that there is a stop sign) approximately 200 meters prior to the intersection. If the driver ignores this information and continues driving then an audible warning can be provided to the driver with just enough time for him or her to resort to panic braking.

The same transponders that would be at the intersection or inside the vehicle could also be sold directly to pedestrians, cyclists, and motorcyclists. As discussed earlier, these transponders are the size

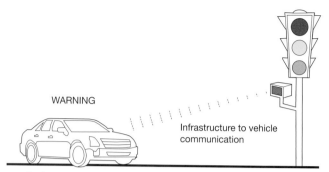

WARNING

Infrastructure to vehicle communication

On-board equipment warns driver if signal violation will occur

Figure 3.3
Infrastructure-to-vehicle (I2V) communications can enhance road safety.

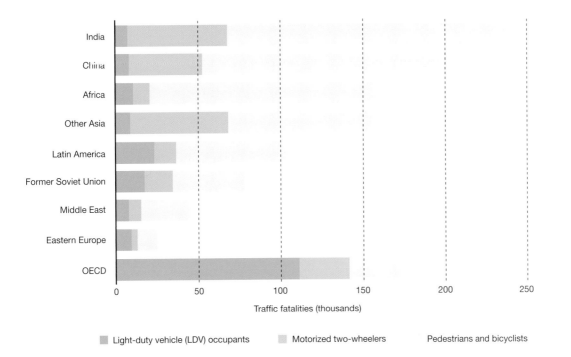

India

China

Africa

Other Asia

Latin America

Former Soviet Union

Middle East

Eastern Europe

OECD

0   50   100   150   200   250

Traffic fatalities (thousands)

Light-duty vehicle (LDV) occupants   Motorized two-wheelers   Pedestrians and bicyclists

Figure 3.4
Traffic-related fatalities for various roads.

of a mobile phone and can communicate their presence to vehicles wirelessly and either eliminate collisions or reduce their severity. This can significantly reduce traffic-related fatalities, because, as figure 3.4 shows, many of these fatalities are vulnerable road users (pedestrians, cyclists, motorcyclists).

Location-Based Services

Networked computing and control can also enable many valuable location-based services. It can, for example, facilitate the search for parking. Sensors in parking spaces can monitor availability and transmit this information to nearby traveling vehicles. This not only reduces driver frustration, but also saves time and energy, and reduces congestion by quickly getting vehicles in search of parking off the road. Furthermore, as we shall see in chapter 8, it provides the basis for sophisticated parking space pricing and allocation schemes.

Different types of sensors in streets and roads can monitor traffic volumes. (This information can also be obtained from GPS navigation systems in vehicles.) This provides the basis for reducing congestion through the mechanism of dynamic road pricing, which will also be covered extensively in chapter 8.

Networked computing and control can also support new, potentially fairer approaches to automobile insurance: with suitable privacy protections in place, rates can be based on tracking of actual driver mileage and behavior. More controversially, such tracking can enable automated traffic law enforcement as well: instances of speeding, running stop lights, and so on can be detected, and fines can be levied automatically.

Still more benefits are obtainable through the electronic integration of different mobility systems. This can, for example, enhance the effectiveness of public transportation systems by making it much easier to plan trips and to coordinate multimodal travel. It allows the commuter to drive to the train station by car while being aware of changing traffic conditions or delays in the train schedule and being able to plan the trip accordingly, which improves time management and reduces stress.

Recapturing the Horse's Intelligence

In chapter 2 we saw that the new DNA would move the automobile further away from the horse-and-carriage architecture and the "horsepower performance" figure of merit. In one important way, however, the reinvented automobile does hark back to the days of horse riding and horse-driven carriages, and that is in its intelligence.

The Westerns that showed the cowboy calling to his horse to pick him up, and then falling onto it and being taken where he needed to go even while asleep in the saddle, present a powerful idea that was lost when the automobile took over. During the twentieth century the driver had to go to the vehicle to get in it, and then had to control and direct the vehicle to its destination. The automobile's new DNA encompasses autonomous operation and will allow a driver to summon the vehicle from its parking spot to pick him or her up. As with the horse, the vehicle could even transport the driver to the desired destination and then go and park by itself.

We often forget how much time is spent looking for parking and then walking back to the parking space afterward.[2] Some studies estimate the average search time at eight minutes in dense, urban environments. This contributes to the relatively slow door-to-door times for car use, making Bus Rapid Transit a faster option, typically, for trips longer than about five miles (figure 3.5). Moreover, a significant proportion of the fuel consumed in congested urban driving conditions is in looking for parking. Connectivity (knowing the nearest parking space availability) and automated parking (leaving the vehicle to drive to the nearest parking space by itself) will combine to save time and energy and provide both personal and societal benefits, which will help to accelerate the commercialization of these vehicles.

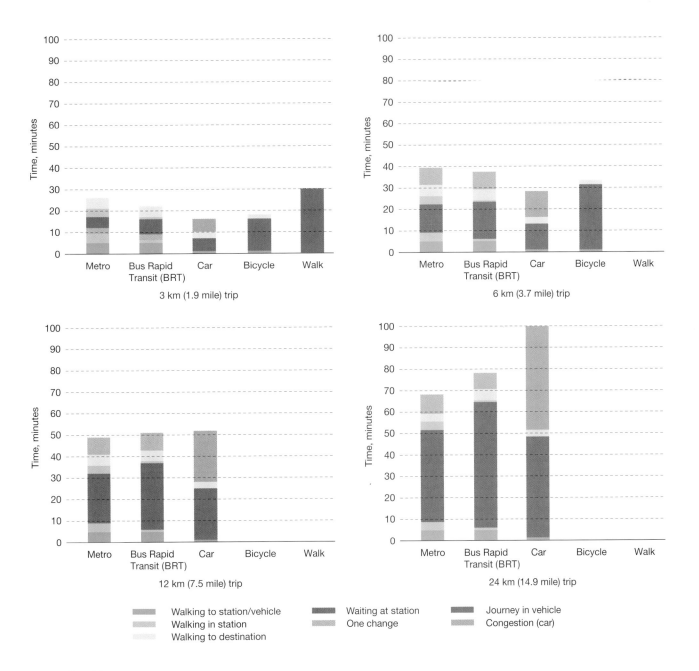

Figure 3.5
Door-to-door times simulated for various forms of transportation.

Electronic Customization

Life is on hold when one is driving. That is what it feels like to many people when they are spending time behind the wheel. This is not true for everyone, of course, and not at all times. But as congestion worsens and as road construction seems more and more like a permanent condition, the fun associated with actually driving a vehicle is decreasing. Many young people value the freedom of mobility, but fewer and fewer are excited by the driving experience itself. Vehicle sales in Japan have been in decline as the population is aging and the youth express little enthusiasm for purchasing a product that spends most of its time parked and most of the rest of the time crawling in traffic. What can we do to enhance the driving experience? Perhaps more importantly, with autonomous operation, what can we do to enhance the ride experience?

Connectivity and intelligence can not only enhance the vehicle's awareness of the external environment, help to optimize traffic flow, and reduce trip times, but also improve driver and passenger interaction with the vehicle through electronic customization. Today, a Bluetooth-enabled telephone can be operated hands-free in the car, but in the near future a driver could take further advantage of wireless remote control by downloading vehicle settings onto a "Superfob" or handheld device, and as the driver approaches the vehicle these personalized selections, such as seat and mirror settings, could be activated. (This capability can be extended to provide automated parking and retrieval.) Chassis-

by-wire technology allows the driver to select brake feel and the steering gain and steering feel, while electric-drive vehicles can be tuned for optimum performance, fuel economy, and even sound signatures or ring tones. Preconditioning of cabin temperature while the vehicle is charging could help to save battery power and range when driving.

Merging Life Inside and Outside the Vehicle

The Superfob could also set the appropriate interior lighting and preload your favorite music, radio, and video content prior to entering the vehicle. This leads to obvious questions, such as:

- Why pay twice for hardware (DVD player both inside the vehicle and at home)?
- Why carry an MP3 player, a keyfob, and a cell phone into the vehicle?
- Why put up with the hassle of bringing content into vehicle from outside (e.g., printing out directions at home)?

We can save money and time by seamlessly integrating connectivity inside and outside the vehicle. In the future, the Superfob will dock into the vehicle's interior and provide all the navigation, music, radio, movie, and Internet content the driver needs.

Furthermore, connectivity will complement mobility by reducing the need for it. Inefficiently looking for parking when connectivity can inform of availability is just one example of this. Another example is driving to a store only to find it is closed or out of the product you want. Connectivity inside the vehicle could help reduce the probability of these pointless journeys, just as people today use computers to shop online.

New Driving and Riding Experiences

In the future, through the use of autonomous driving technology (as discussed in chapter 2), automobile users will be able to choose between driving and riding. Both experiences will be improved. If the driver chooses to "grab hold of the wheel" then he or she will experience a vehicle that has lots of torque at low speeds, allowing the vehicle to accelerate like a jet. Motors in the wheels can also provide the vehicle with the ability to crawl sideways or turn sharply at the last moment at intersections, or to turn on a dime—making reversing out of a parking spot a thing of the past (more on this in chapter 4). The steering and brake feel can be personalized based on driver preferences and road conditions, so that, for example, vibrations through the steering column are eliminated when going over a bumpy road. Just as people pay to download personalized ringtones onto their cell phones, it will be possible to personalize the sound of electric-drive since there is no engine growl to compete with.

What if the driver chooses "autopilot" mode? Providing that the vehicle's operating environment is appropriately enabled, this will be possible. Remember the Superfob that docked into the vehicle, bringing infotainment content with it. When one is driving, information such as navigation display, range to destination, driving speed, location of nearest charging ports, and so on could be displayed. When one is riding, this information could be moved to the background so that other information can be presented. If the "driver" wants to relax then he or she could surf the Internet or watch a movie or a Tivo-ed recording. Or maybe the driver wants to use commuting time to catch up on work by calling some business colleagues or making last-minute changes to a presentation she must give that morning at work. Or maybe the priority is to grab a quick lunch, or catch up on some sleep. The experience would be like being in a taxi—except there is no driver.

Figure 3.6 illustrates some of the new driver experiences enabled by autonomous operation. On the left, control is from the left seat. In the middle, operation is hands-free. On the right, control is from the right seat.

Figure 3.6
New driving experiences enabled by autonomous operation. Control shifts from left to right as required.

Some categories of drivers—those who are alcohol-impaired, those who have disabilities that limit their driving capabilities, and the elderly with slow reaction times, for example—should *not* grab the wheel. Currently, their only options are to accept limitations on their personal mobility or to create danger to themselves and others by driving anyway. Another benefit of autonomous driving, then, is safe extension of personal mobility to these drivers.

Social Networking on Wheels

From a social perspective, a traditional automobile is an isolated room on wheels. It enables conversation and social interaction within its walls, but isolates the enclosed traveling group from others. In other words, it rigidly and arbitrarily modularizes social grouping.

Now consider a typical family of two parents and two children. Imagine they have limited income and can, right now, only afford one vehicle. For this family, only one adult has use of the vehicle at a time during the week, while the other will need to find other forms of mobility (walking, cycling, ride-sharing, taxi, bus, train, and so on). The person who has the use of the vehicle has a larger vehicle than he or she probably needs, since it can accommodate at least four people and is probably only transporting one. This translates into higher running costs for fuel and, perhaps, greater difficulty finding an available parking space. A larger, heavier vehicle also places a bigger burden on society by producing more greenhouse gas emissions, consuming more oil, and posing more dangers to vulnerable road users, such as pedestrians and cyclists.

Imagine, now, how the Mobility Internet can change this. Each parent owns a small two-seat battery-electric vehicle (as described in chapter 4), which together could cost about the same as one traditional car. Each parent now has mobility during the week that is appropriate for commuting with some utility in reserve for picking up an unexpected person or for carrying goods from the supermarket. These individual vehicles are each easier to park than one large vehicle. On the weekend or during the evenings when the family may want to go out together, these two vehicles will be wirelessly connected with each other and will travel as a virtual train, occupying a footprint similar to that of a large SUV (think two very small cars with a short separation or platooning distance between them). These two vehicles are not only capable of traveling together but one can "peel" off from the other if need be (unlike the current scenario, where all four family members must go everywhere together). What about social interaction between the four family members? It could be improved. With a web camera and a display inside each vehicle it would be possible for all four members to be able to talk face to face with each other, something that is not possible today since the driver has to face forward when driving.

This creates limitless potential for spontaneous connections with anyone on the road at the same time. You could be driving around and realize that one of your friends is nearer than you thought, and you could videoconference them and arrange to link up and go somewhere together. This is the mobility equivalent of the pop-up bar that tells you when one of your "buddies" is online.

Summary: A Revolution in Movement and Interaction

Treating an automobile as a stand-alone product neglects many of the externalities associated with its actual operation. Unlike a refrigerator, for example, which has a clearly defined energy-efficiency rating, the efficiency of an automobile depends on how close it is to other vehicles, since the energy efficiency when operating on the open road can be much higher than when stuck in traffic. Connectivity is the key to coordinating automobile movement and to optimizing the flows of large numbers of vehicles in confined spaces to reduce energy consumption, pollutant emissions, bottlenecks, and accidents.

Connectivity will also release the most precious commodity of all to many people—time. When drivers are given appropriate and timely traffic information, for example, they will be able to make their trips shorter, more predictable, and less stressful. The value proposition is magnified further when the vehicle is capable of autonomous operation, as this will allow drivers to disengage from driving and let them do pretty much what they want. This could be resting, working, relaxing, entertaining themselves, or social networking.

Renewal of transportation infrastructure to enable convenient, efficient, sustainable personal mo-bility in the twenty-first century is not, then, simply a matter of repairing roads and bridges and providing more traffic lanes. It requires, instead, the addition of some fundamentally new capabilities—those of the Mobility Internet.

From a business perspective, the Mobility Internet opens up some interesting possibilities. In many ways, small, intelligent electric vehicles are more like networked consumer electronics devices—as represented by laptop computers, smart phones, and iPods—than traditional automobiles. They are small, light, and comparatively inexpensive (though still with the potential to be profitable), and they have high electronics and software content. Through the Mobility Internet, they provide access to many valuable, computationally based services: smoothing of traffic flow; avoidance of accidents; efficient, reliable navigation to destinations; location of parking spaces; supply of vehicles wherever and whenever they are needed (in the mobility-on-demand services that will be described in chapter 8); urban guidance, commentary, and location-based advertising; and many more. In addition to providing access to these basic mobility services, light electric automobiles connected to the Mobility Internet will—much like today's smart phones—serve as platforms for all manner of "apps" provided by innovative third-party developers.

Nobody could predict exactly all the ways in which the Internet would be used. Similarly, we can be sure that people will value the freedom that the Mobility Internet provides and will invent new ways to take advantage of it to enhance their lives.

4

Reinventing the Automobile
for Urban Use

As we have already pointed out, a typical automobile today is larger and heavier than it needs to be to provide personal urban mobility, and its capabilities are excessive for this purpose. We have noted that it weighs about 20 times as much as its driver, can travel over 300 miles without refueling, and can attain speeds greater than 100 miles per hour. It requires more than 100 square feet of valuable urban real estate for parking, and it is parked about 90 percent of the time.

All these forms of inefficiency and lost opportunity are multiplied, to a staggering degree, by the enormous numbers of cars on the road today. In the United States alone, there are now 250 million cars and trucks traveling three trillion miles each year on four million miles of roads, utilizing 170,000 gas stations. Their internal combustion engines account for about one-third of the annual national energy consumption.[1]

Furthermore, vehicle usage inefficiencies are only the tip of the iceberg. Automobiles function as elements of large-scale, complex systems that include streets and roads, parking, energy supply and waste-removal networks, policies and regulations, and associated enterprises.

The various subsystems have coevolved: automobiles have adapted to the forms and dimensions of streets, but streets have also adapted to the capabilities of automobiles; automobiles enable suburban sprawl, while suburban sprawl creates demand for automobiles. These codependencies magnify certain negative impacts of today's automobiles and create some barriers to change, but they also promise to scale up the positive effects of improvements.

Can intelligent, networked, electric vehicles—utilizing the technologies and design strategies that we have described—achieve a better match between capabilities and need? The answer is yes. In this chapter we will explore the design principles and features of such vehicles. In essence, future urban automobiles will need to be less SUV and more USV—for Ultra Small Vehicle. In chapter 9 we will go on to demonstrate how networks of USVs can achieve very significant benefits for consumers and for urban societies.

Limitations of Existing Vehicle Designs for Urban Use

Each vehicle design represents a trade-off involving multiple requirements and constraints, provides a different value proposition to the customer, and has a different impact on the urban environment. We can see this in the differences between today's pickup trucks, crossovers, sedans, and sports cars. Each has a very different level of utility and performance, but all offer high-speed, long-distance operation suitable for intercity use. Figure 4.1 shows a subjective comparison of various vehicles' urban mobility applications, ranging from a bicycle to a conventional automobile. Even a small battery-electric city car (of traditional design) is still relatively heavy and expensive because it is typically designed to operate on highways as well and must, therefore,

be capable of withstanding an impact at speeds of 35 mph, be able to travel at speeds above 75 mph, and be able to drive for distances of around 100 miles. Moreover, a small battery-electric car, as we will see shortly, is typically about twice as large and twice as heavy as needed for urban use, which has serious implications for its parking footprint and its safety for vulnerable road users.

A neighborhood electric vehicle (NEV) comes close to achieving a reasonable balance for urban use. However, it too retains the DNA of the conventional automobile. NEVs are, effectively, stripped-down conventional cars with electric-drive systems. The fact that they operate as stand-alone vehicles limits what they can do to address the societal issues of safety and congestion. Although reduced in size, they are still roughly twice as large as necessary, which again affects their parking requirements. Moreover, the consumer benefits associated with the design renaissance of the new DNA—such as greater maneuverability, more pleasurable driving experience, greater ease of entry and exit, tailored driver controls, automated parking, and personal connectivity—are missing, and this further limits consumer appeal and the potential mass market.

Potential for a New Type of 100-Inch, 1,000-Pound Vehicle: The USV

So what might a USV, designed for urban use, look like?

A first consideration is the number and arrangement of seats. A one-seat pod occupies minimal space,

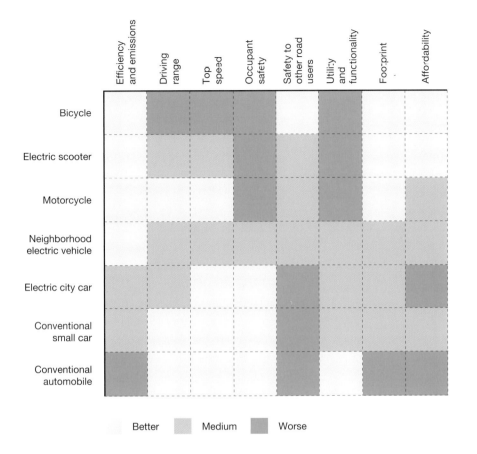

Figure 4.1
A subjective comparison of various personal urban mobility options.

uses minimal energy, and has minimal cost, but also has limited utility and flexibility. On the other hand, adding three or four seats leads to a much larger vehicle with its associated trade-offs in terms of size, mass, and cost. Given that the average occupancy rates in automobiles used in urban environments are between one and two and that there is limited land space available in a city, a good starting point is to design a two-seat vehicle. This is probably the "sweet spot" between urban-friendly and customer-friendly. Because people are social creatures, the two seats should be side by side to encourage face-to-face interaction and to make it easier to communicate.

Although bicycles and scooters are popular in many cities around the world, there is a universal desire to own an automobile as soon as it becomes affordable. No doubt a part of this is the status that being a car-owner confers, but the automobile also possesses some significant functional attributes that people desire. We like to be comfortable, have some privacy, and be safe, and the automobile provides all these benefits by protecting the occupants from the outside environment and from other road users. Automobiles also provide some storage flexibility for carrying an extra person and for carrying items such as shopping bags.

In designing an urban vehicle, we need to retain these key features. But we can do so in far more economical fashion than traditional automobiles have led us to expect. Average traffic speeds in cities (below 20 mph) and typical daily driving distances for urbanites (less than 25 miles) provide significant opportunity for mass and cost reduction, compared with conventional vehicles that can travel 300 miles nonstop and can exceed speeds of 100 mph. It is these latter requirements that substantially drive up the cost of the vehicle (through larger, heavier, more costly propulsion and chassis systems and in the supporting body structure); and it is by reducing these requirements that we can begin to design a much friendlier vehicle for the city and its residents.

Starting with a two-seater, it is entirely conceivable that vehicles can weigh less than 1,000 pounds and be shorter than 100 inches in length, making them three to four times less mass- and space-intensive than conventional automobiles. How can a USV be designed to have a minimum footprint when it is parked (which is most of the time)? Two possible approaches (and there are likely to be many others) are through folding, as demonstrated by MIT Media Lab's CityCar, and through dynamic stabilization, as shown in GM's Project P.U.M.A.[2] Despite some differences, these two approaches share many elements of the new DNA and tailor it to urban applications.

The Simplicity of Battery-Electric Vehicles

As discussed in chapter 2, the battery-electric is probably the best electric-drive approach for urban mobility applications, where modest range and performance requirements are acceptable and where zero emissions and compact packaging are most needed.

To paraphrase Einstein, "The best design is the simplest one that works." The degree to which battery-electric-drive systems can simplify an automobile, especially if a clean-sheet approach is taken, is truly profound. In a traditional car, fuel lines connect a tank to a large internal combustion engine, an exhaust system carries combustion products to the tailpipe, and a mechanical drive train connects the engine to the wheels to provide propulsion. In a hybrid car, the propulsion system is even more complex, since there are batteries in addition to a tank, and there is both a gasoline engine and an electric motor. In a battery-electric-drive vehicle, though, much of this mechanical complexity can be eliminated: essentially, there are just batteries, wires, power and control electronics, and wheels. And the complexity that remains might even be encapsulated within a small zone of the vehicle, especially if the electric motors are located right at the wheel hubs.

Furthermore, there are fewer different items to contain and move around in an electric-drive vehicle, and fewer associated technologies to integrate and manage. A traditional car requires elaborate systems of reservoirs, tubes, valves, and pumps to distribute the gasoline, oil, water, air, and exhaust gases, but a battery-electric automobile replaces most of these complicated distribution systems with wires connecting the batteries to the wheels. This simplifies design, manufacture, and maintenance. It reduces overall mass and cost. And at the end of the vehicle's life it simplifies disassembly and recycling.

There is a close analogy, here, with the clarification and simplification of product designs that resulted from the development of digital technology in the second half of the twentieth century. Not so long ago, separate analog channels distributed voice, video, text, and numerical data to different types of user devices—telephones, televisions, and so on. Today, though, digital channels carry bits to digital devices that are all much the same inside. The digital revolution produced powerful technological convergence and also simplified design and manufacturing tasks, by converting all forms of information to bits. In the early twenty-first century, the conversion of all forms of energy to electricity will have a similar effect.

From a vehicle design perspective, the packaging freedom associated with both the skateboard chassis (no engine compartment) and crash elimination (no front-end crumple zone) can help to make vehicles dramatically shorter, which is particularly useful where parking space is limited and expensive. It is quite conceivable that future two-seaters could be half as long as the smallest cars or NEVs today.

Wheel Motors

Wheel motors have a significant impact on the architecture of the USV. Space is critical in an urban setting, and wheel motors allow the vehicle to be shrunk by moving the motors outboard into the wheels.

With conventional electric motor designs (a single electric motor on the axle and half shafts leading to each wheel) the end result is a package similar to that of a gasoline-powered car. Elegantly packaging batteries is a challenge in electric-drive vehicles with single electric motors as they are bulky and require cooling, and the traditional automobile architecture evolved without considering or providing for them.

As a result, when batteries are simply inserted into conventional automobile forms that have been designed around the internal combustion engine, they usually end up being squeezed uncomfortably under the rear seat, between the seats, or occupying what would otherwise be trunk space. But cars with modular in-wheel motors provide opportunities for better battery packaging because they free up extra space under the floor or in the front compartment. This extra battery space can also provide vehicles with greater range and performance.

Wheel motors also reduce the space needed to maneuver the vehicle. They enable greater vehicle accessibility in constricted urban contexts such as those found in European cities, and they significantly reduce carriageway and parking space requirements. A reduced turning radius means that the car can negotiate sharper corners, and "O-turn" capability means that dead ends present much less of a problem. This type of omnidirectional movement can make it much easier to maneuver out of a driveway or to "reverse" downhill in icy conditions. Together with by-wire systems, wheel motors let each wheel be independently controllable, thus offering all-wheel steering, stability, and traction control almost "for free." Together with the space-saving attributes, all these features make wheel motors a very attractive solution for urban mobility.

Despite these advantages and some familiarity (Ferdinand Porsche designed a battery-electric car with four in-wheel motors in 1900, and wheel motors have commonly been used in trains and heavy equipment), wheel motors have often been dismissed

on two counts: the additional cost of extra motors and controllers and the increased unsprung mass (mass that is not supported by springs or the suspension system, such as wheels, tires, and suspension—higher unsprung mass leads to worse ride and handling quality). It should be possible to mitigate the effects of unsprung mass through a combination of clever design, lighter materials, and appropriate suspension tuning, and parts simplification could create substantial cost savings in manufacturing assembly.

Robot Wheels

The advantages of wheel motors would be limited, however, if they were simply added with little optimization and few modifications to other vehicle systems. If, on the other hand, we design the vehicle around the wheel motors' full capability, then it should be possible to compensate for their additional mass and cost by simplifying other systems. Michelin, for example, has developed an "Active Wheel System" that integrates into the space inside the wheel a motor for propulsion, an active suspension for ride, handling, and comfort, and a standard disc brake for friction braking. The ability of this system to transmit different amounts of torque at each wheel can create a highly sophisticated differential, which allows for much improved handling in turns.

It may be possible to go even further. Wheel motors, like all electric motors, can also act as generators: they can recover braking energy that would otherwise be lost and feed the current back into the battery (so-called regenerative braking). It should be possible to reduce or even eliminate the need for friction brakes, which will reduce the cost and unsprung mass penalty. Wheel motors at all four corners can recover more braking energy than a single electric motor, which can help reduce the size of the battery or fuel cell for a given range by more than 10 percent, yielding further mass and cost savings.

Over the last thirty years or so, chassis system functionality has become ever more complex with each successive vehicle generation, as new electronic features are added on to the conventional vehicle's mechanical underpinning. There comes a point, however, where it makes sense to start with a clean sheet and begin anew with the concept of providing and controlling torque at each corner of the vehicle. If wheel motors are adopted, then many chassis and chassis electronics systems (brakes, antilock brakes, traction control, electronic stability control, all-wheel drive, electric power steering, four-wheel steering, torque vectoring, and so on) can be eliminated or downsized. Wheel motors may even exceed the capability of conventional chassis systems because more precise control becomes possible with high-resolution sensing of the wheel motor position and the ability to change its direction of rotation or to stop it very rapidly.

Ultimately, the combination of wheel motors and by-wire systems will allow each corner of the vehicle to be electrically powered and digitally controlled. These modules will provide propulsion, braking, suspension, and steering and can be designed as modular snap-on units, like USB devices for personal computers or bayonet-mount lenses for cameras. As with USB devices, they can have standard interfaces—providing structural, electrical, and data connection to the chassis. Thus they become self-contained "robot wheels."

From a manufacturing perspective, these modular units—like disk drives in laptop computers—have the advantage of encapsulating multiple functions and a good deal of mechanical and electronic complexity behind their standard interfaces. Overall, modular electric-drive cars with in-wheel motors have far fewer parts, fewer subsystems, and fewer interfaces than gasoline cars. Their advantage over complex hybrid electric cars is even greater. This simplifies supply chains and assembly processes, and it enables competitive innovation on the wheel side of the interface to enhance performance and to drive down costs.

From a repair and maintenance perspective, using in-wheel motors keeps cars out of the repair shop and reduces the need for skilled mechanics. If a robot-wheel unit breaks down or reaches the end of its useful life, it can simply be snapped out and replaced with a new one. This is an operation that is quick and easy and can be performed anywhere. The old wheel can either be sent to a central repair shop for offline repair or consigned to recycling. Furthermore, wheel-motor cars not only have fewer parts and simpler mechanical organization than gasoline and hybrid cars, they also have around one-tenth as many *moving* parts (which reduces wear, increases reliability, and reduces maintenance requirements over vehicle lifetimes).

Emerging Vehicle Concepts

One of the great advantages of in-wheel-motor electric-drive vehicles is that they allow a wide variety of vehicle configurations. These configurations can respond to different requirements, conditions, and decisions about technical trade-offs. In recent years, for example, several personal mobility and small city car concepts have been demonstrated, by Japanese automakers in particular. These exploit wheel motors because they provide clean mobility in vehicle designs that need to be compact and maneuverable (figure 4.2). They are mostly intended for use by one person only, over very short distances, in dense urban environments.

The USVs described in the following sections fill a different niche in the urban mobility ecosystem. They transport two people; they travel longer distances before recharging or refueling than these personal mobility devices; and they have higher top speeds. However, since they are intended specifically for urban use, they are not designed to match the passenger and baggage capacity, range, or speed of today's typical automobiles. They do have some capacity to operate on highways outside urban centers.

In other words, they represent a carefully chosen trade-off point. They have more utility than single-passenger personal utility devices, but higher mass, energy consumption, and cost. On the other hand, they have much lower mass, energy consumption, and cost than today's automobiles.

GM's P.U.M.A. develops the possibilities of the USV in one way. The MIT Media Lab's CityCar (figure 4.3), which we shall now describe, develops them in another.

Figure 4.2
Personal Mobility concepts.

Toyota Winglet

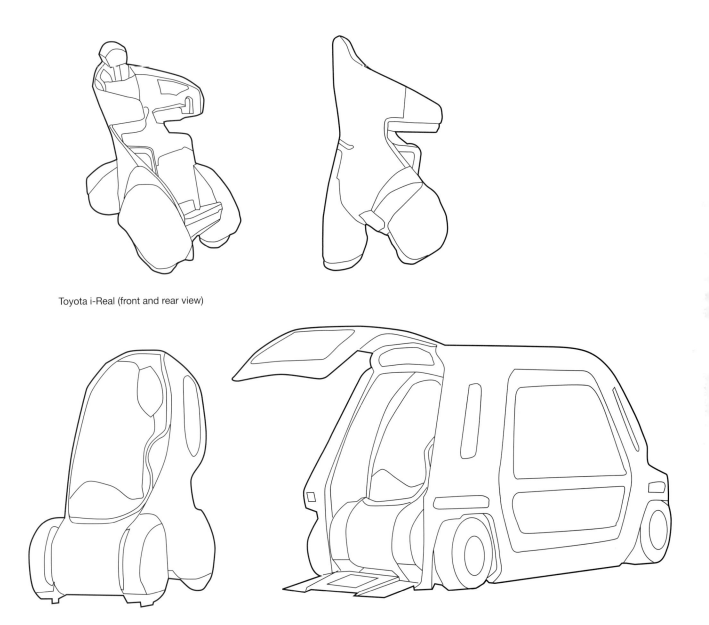

Toyota i-Real (front and rear view)

Suzuki SSC + PIXY (with PIXY inside and outside SSC)

Figure 4.3
The MIT Media Lab's CityCar.

The MIT Media Laboratory's CityCar

Figure 4.4 illustrates the basic wheel layout possibilities for in-wheel-motor vehicles with up to four wheels.

One-wheelers—much like unicycles—are a logical possibility, but so far have not found practical application in mobility systems.

Two-wheelers can take the familiar form of bicycles, motor scooters, and motorbikes, in which the rider provides the necessary balancing. MIT's Smart Cities group has also developed and prototyped the RoboScooter—a folding electric motor scooter with in-wheel motors. The alternative of side-by-side wheels is represented by Segways and P.U.M.A. vehicles, in which balancing is provided by an electronically controlled dynamic stabilization mechanism. These are steered not by turning their wheels, but by rotating their wheels at different speeds.

Three-wheeled vehicles can have the single wheel at the front or the back and can also be structured as circular vehicles that don't have fronts or backs.

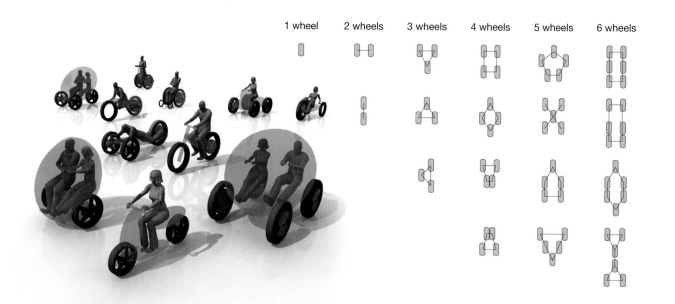

Figure 4.4
Wheel layout possibilities for in-wheel-motor vehicles.

Three-wheelers can have stability problems, so they have seen only occasional use in North America and Europe, but they have been popular in many crowded Asian cities—as, for example, in Indian auto rickshaws—where very low speeds reduce the importance of stability considerations. Four wheels can be arranged in the familiar rectangular configuration, or in a diamond pattern—as employed, for example, in the ITRI LEV (Light Electric Vehicle) concept vehicle, which premiered at the 2007 Milan Motorcycle Exhibition. Five-wheel and six-wheel vehicles have some conceptual interest, but are unlikely to be economical in this application.

The CityCar is designed for somewhat higher speeds and longer distances than the P.U.M.A. vehicles (though still less than the speeds and distances of a traditional gasoline-powered car), so it is built around a different trade-off point. In the interests of providing stability in a simple, effective way without expending energy on balancing, the CityCar adopts a standard four-corner wheel configuration.

Each wheel is independently, digitally controlled, and can vary both direction and speed. This allows a wide range of maneuvers, making the CityCar particularly suitable for tight urban conditions (figure 4.5). In addition to conventional steering (left side of figure 4.5), wheel motors provide various four-wheel steering capabilities. CityCars can, for example, execute sideways or crablike motions for parallel parking (right side of figure 4.5), and O-turns in place of three-point (or five-point!) turns, as shown in the center of figure 4.5.

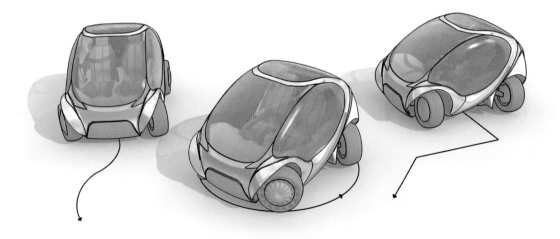

Figure 4.5
Some CityCar movements enabled by its independently controlled wheel motors.

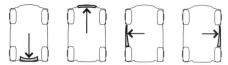

1-door configurations

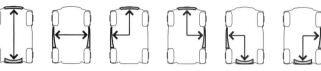

2-door configurations

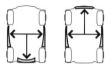

3-door configurations

4-door configurations

Figure 4.6

Entry and exit possibilities for a four-wheeler with wheel motors. The CityCar provides front entry and exit for the passenger compartment combined with rear entry and exit for the baggage compartment.

This wheel system could also provide redundancy so that if any one wheel fails, or even two wheels fail simultaneously, the CityCar could continue to drive safely on its remaining wheels.

This arrangement potentially allows entry and exit, between the wheels, at four points around the perimeter of the vehicle (figure 4.6). In addition to the usual possibilities, front entry and exit becomes feasible because there is no engine in the way. In the CityCar, passenger entry and exit is at the front, baggage and emergency entry and exit are at the rear, and there is no side entry and exit. This allows nose-in parking to the curb and passenger embarkation and disembarkation from the sidewalk rather than from the road, and eliminates most of the need for side clearance between parked automobiles. It

also simplifies the design of the sides of the vehicle, which need not accommodate door openings. It may also permit a more robust side structure to enhance side-impact performance.

By means of a four-bar linkage, the CityCar folds up for more compact parking (figure 4.7). This adds some weight and complexity, of course, but it enables a vehicle with an extended wheelbase and low center of mass for driving, combined with minimal footprint and ease of ingress and egress for parking. In cities where parking space is scarce and expensive, this could represent an attractive trade-off.

The CityCar has a completely digital, drive-by-wire driver interface. It is driven with a two-handed joystick (figure 4.8). The driver pushes the handles forward to accelerate, pulls them back to brake, and

Figure 4.7
Folding the CityCar.

Figure 4.8
The CityCar's two-handed joystick driver interface.

rotates them to steer. A flat video screen on the front door provides dashboard information. This arrangement provides an extremely simple interface, keeps the interior very clean, and creates no obstructions to front entry and exit.

The baggage compartment is located on the other side of the folding mechanism from the passenger compartment (figure 4.9). This allows it to remain low and horizontal when the vehicle is folded. Batteries, a heavy part of the vehicle, are located in the floor. This keeps the vehicle's center of mass low, even when folded, and allows for battery cooling.

The battery-charging point is located under the vehicle. This allows the use of a "smart curb" charging device, which can operate either by contact or inductively (see chapter 6). Thus the CityCar can be plugged in, but it does not have to be: it can be connected automatically to the grid whenever it is parked in a suitably equipped parking bay.

Construction of the CityCar is simple and modular, and the number of parts is very low compared to a gasoline-powered automobile or a hybrid (figure 4.10). The CityCar eliminates sheet metal, paint, and many of the complex details traditionally found in automobiles. It could have a rigid, cast aluminum exoskeleton and polycarbonate panels— similar to the cockpits of fighter planes. The panels snap in and out, and the side panels can be removed for emergency exit.

Safety systems operate at multiple levels. As with the P.U.M.A. vehicles, electronic sensing and wireless communications can be employed to greatly reduce the likelihood of crashes. If crashes do occur, the low

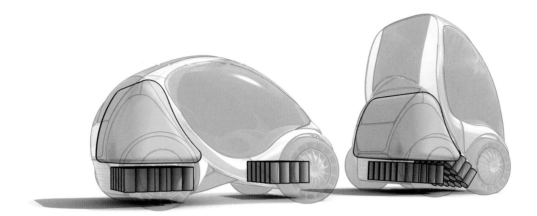

Figure 4.9
Battery and baggage storage in the CityCar.

Figure 4.10
Simple, modular construction of the CityCar.

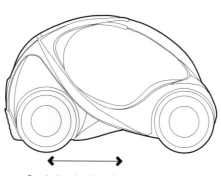
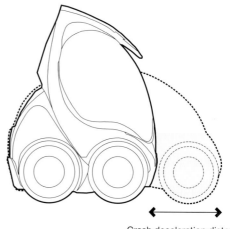

Crash deceleration distance

Crash deceleration distance

Figure 4.11
The CityCar's folding mechanism also provides space
for crash-deceleration systems.

mass and relatively low speed can greatly reduce the energy of a crash with similar vehicles. Seat belts would restrain passengers in the usual way, and airbags may still be needed for the CityCar because of its somewhat higher operating speed than the P.U.M.A. vehicles.

There are no traditional front- and rear-crush zones to provide controlled deceleration of the passenger cabin in the event of crashes. (A crush zone is the space required for absorbing impact energy from collisions.) These, of course, would add length to a vehicle. Additional safety equipment will be needed to compensate for the rigid body, but, as with folding, this may be a desirable trade-off in the interests of compactness. There are also opportunities to provide controlled deceleration through a crush zone or high-speed shock absorbers integrated with the

folding mechanism (figure 4.11). On front or rear impact, the CityCar absorbs evergy by folding up at a controlled rate.

Finally, the CityCar's modular construction, together with its replacement of traditional automobile hardware by electronics and software, opens up many opportunities for design variation and customization within the same fundamental architecture (figure 4.12). Wheel units can be upgraded, just as disk drives can be upgraded in laptop computers. Within the structural exoskeleton, side and front panels and rear luggage compartments can be varied freely—either before or after sale. Taking advantage of these features, CityCars are designed to be configured and customized online by purchasers, much like today's personal computers or sneakers.

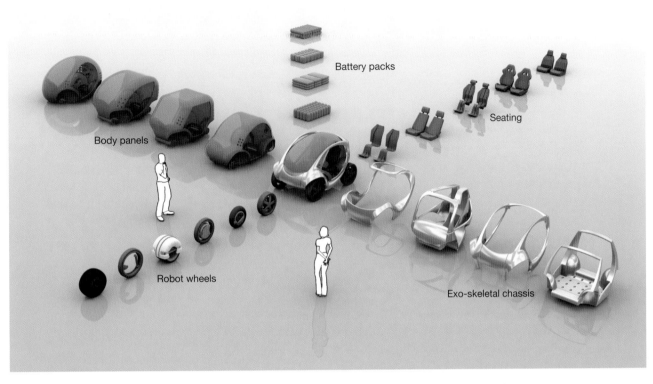

Body panels

Robot wheels

Battery packs

Seating

Exo-skeletal chassis

Figure 4.12
Customization opportunities with CityCars.

A further, very dynamic level of customization is provided through software. Many driving characteristics of the CityCar are embodied in the software of the digital control system rather than in hardware (which is difficult to alter), a very powerful feature. When the vehicle identifies its driver, it can load preferences that are stored onboard or downloaded wirelessly from a server. Whatever vehicle you drive—your own, borrowed, or one in a shared-use system—it will always feel like *your* vehicle. Furthermore, with the provision of application programmer interfaces, the CityCar can—as demonstrated by Apple's iPhone—become a platform for lively third-party software innovation that can rapidly extend its capabilities.

Project P.U.M.A. (Personal Urban Mobility and Accessibility)

The CityCar represents just one of many attractive combinations of features and trade-offs. Let us now consider another.

In compact, two-seater, front entry/exit cars with in-wheel motors, there is another way to reduce footprint: by eliminating the rear wheels. This yields a car that is not much longer than a single wheel and has a width determined by seating requirements, much like a rickshaw cabin. This vehicle design takes the maxim "Man Maximum, Machine Minimum" seriously.

The rear wheels, of course, are normally needed for stability, but the Segway Personal Transporter (PT) has demonstrated the feasibility of maintaining balance electronically, and the principle may be extended to larger vehicles. It represents a different design trade-off from that illustrated by folding cars, because it eliminates the complexity of the folding mechanism but instead introduces the complexity of electronic balancing. It is extremely compact, both parked and in motion.

The Segway PT has shown that, at least for very lightweight and low-speed battery-electric vehicles, it is possible to eliminate conventional brakes, steering, and even liquid cooling systems with wheel motors. A schematic of its chassis, shown in figure 4.13, illustrates the simplicity of its "skateboard." Braking is provided by the electric motors acting as generators, and even when the battery is fully charged, braking still occurs through self-heating of the motors. The PT weighs around 100 pounds and has a governed top speed of 12.5 mph, but it may be possible to extend this paradigm to a larger vehicle having a mass of around 700–800 pounds and a top speed of 25–35 mph, which is still far short of a conventional automobile (3,000 pounds and 100 mph).

In April 2009, General Motors and Segway introduced to the public a proof of concept for such a type of vehicle, dubbed Project P.U.M.A., as shown in figure 4.14. This vehicle prototype is a battery-electric vehicle with two seats side by side and with just two wheels side by side. The prototype shown can travel at speeds of over 25 miles per hour and has a range of around 25 miles between recharges.

The Project P.U.M.A. vehicle combines several technologies that were previously demonstrated by

Chapter 4

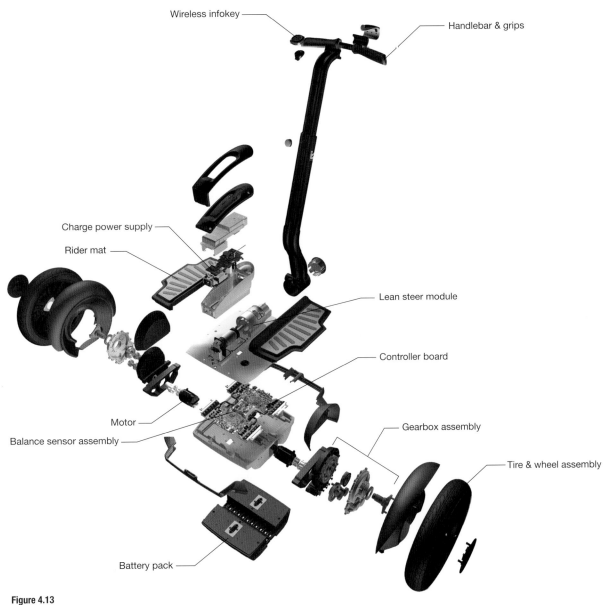

Wireless infokey

Handlebar & grips

Charge power supply

Rider mat

Lean steer module

Controller board

Motor

Gearbox assembly

Balance sensor assembly

Tire & wheel assembly

Battery pack

Figure 4.13
Exploded view of the Segway PT.

Figure 4.14
The Project P.U.M.A. prototype, introduced
in New York in April 2009.

Figure 4.15
Dynamically stabilized P.U.M.A.

Figure 4.16
The P.U.M.A. wheel configuration allows convenient passenger entry and exit from the front.

either General Motors or Segway: lithium-ion battery-power; wheel motors to provide by-wire acceleration, braking, and steering; dynamic stabilization (two-wheel balancing); vehicle-to-vehicle communications; dockable handheld user interface, allowing off-board connectivity; and autonomous driving and parking. These technologies are integrated together in Project P.U.M.A. to enhance freedom of movement, while also providing energy efficiency, zero emissions, a high level of safety, seamless connectivity, easy parking, and reduced congestion in cities. A side profile of the exterior design of a "generic" P.U.M.A. vehicle is shown in figure 4.15.

The vehicle is fun to drive, as it is highly maneuverable and has instantaneous torque from the moment of launch. The dynamically stabilized platform enables new ways to enter the vehicle (figure 4.16)

Chapter 4

Figure 4.17
Various P.U.M.A. designs.

and new design forms that can be highly varied. The benefit of two wheels side by side is clear from the designs shown in figure 4.17. P.U.M.A. vehicle designs can enable exterior design variation not only because of the skateboard chassis but also because the wheels do not physically turn from side to side, as on a typical vehicle. On P.U.M.A. vehicles the steering is achieved by requesting different amounts of torque for each wheel, which makes the vehicle pivot and change direction. This allows the option of skirted wheels, creating a very different look. The dynamically stabilized platform also enables new ways to control the vehicle. Dynamic stabilization also manifests itself in the vehicle's quasi-biological expression of movement. For example, when the vehicle is parked it is docked on the front landing wheels to ensure that no energy is consumed by balancing. As the driver approaches or gets into the vehicle, it wakes up, and the vehicle lifts off the landing wheels and balances on the drivable wheels. If the driver is stationary for short periods, the vehicle appears to "dither" as the wheels slide gently and smoothly fore and aft to maintain balance. Dynamic stabilization is maintained with an array of angular rate sensors and accelerometers that determine the orientation and motion of the vehicle's platform. The balance system senses and requests power to the motors to drive the wheels forward or back, depending on whether the vehicle is accelerating or braking. This movement of the wheels relative to the body gives it a kinetic appearance. And when the vehicle approaches an intersection and comes to a stop, it drops down onto the front landing wheels,

Figure 4.18
USVs, suitable for shared use, that fold and stack to achieve a very small parking footprint. (The bitCar concept by Franco Vairani.)

Figure 4.19
In shared-use systems, folding and stacking USVs can be dispensed from the fronts of stacks and returned to the backs of stacks, enabling new types of parking structures.

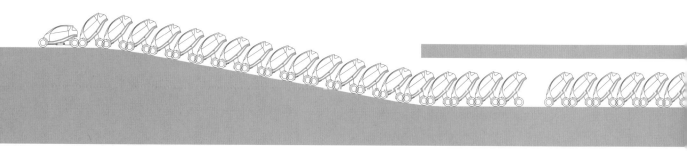

in a process reminiscent of bowing to the other vehicles, pedestrians, and cyclists. When the vehicle is in a tight spot it can spin on its own axis. The biological metaphor is completed when one recalls that humans are two legged and balance in a similar way to the P.U.M.A. vehicles.

With both the P.U.M.A. and the CityCar vehicle approach (and other USV solutions, such as the "folding and stacking" solutions shown in figures 4.18 and 4.19[3]) there is a significant reduction in parking space requirements, and not only because the vehicle has a dramatically smaller footprint. The P.U.M.A. reduces both length and width through its use of only two wheels; the larger CityCar reduces to a comparable parking length by folding; and stacking shared-use USVs overlap when parked (like shopping carts) to eliminate space normally devoted to clearances and access. The high maneuverability of all these vehicles also cuts down on the requirement for access lanes and backing and turning space in parking lots and parking structures. There is also the potential to incorporate some means of automated parking to eliminate the required space for opening and closing doors and to prevent damage when parked. This combination of small footprint, high maneuverability, and automated parking could reduce gross parking space requirements by a factor of four or more (as discussed in chapter 9).

Affordability

These types of USVs should also be significantly less expensive than conventional automobiles. Many people think of battery-electric vehicles as being expensive, but cost is driven by vehicle requirements. When a conventional vehicle is retrofitted with a battery-powered system that must provide similar performance (~100 mph top speed) and significant range (>100 miles), then the upfront cost of a battery-electric vehicle is often increased; but note that millions of battery-electric vehicles are sold each year for far less than $1,000. They are called electric bikes, and they have 25 miles range and 25 mph top speed—and 16 million were sold in 2008 in China alone. Figure 4.20 provides a comparison between various types of bikes and automobiles in terms of performance, vehicle, and energy costs.

USVs will be more expensive than electric bikes, but they should be much less costly than conventional cars (figure 4.21). They will weigh under 1,000 pounds, and if they are powered by a 4-kilowatt-hour lithium-ion battery pack and propelled with two 5-kilowatt wheel motors, they should have sufficient performance to meet the range and speed needs of urban drivers.

Consumers, understandably, tend to focus on the highly visible vehicle-purchase and running-energy

Figure 4.20

A comparison of some personal mobility products for urban use.

	Approx. cost ($)	Vehicle mass (kg)	Maximum power (kW)	Top seed (mph)	Range (miles)	Energy consumption to drive 20 miles (kWh)	Energy cost to drive 10,000 mi/yr ($) (assume 10 cents/kWh electricity and $3/gallon gasoline)
Bicycle (1 rider)	<100	15	0.2	15	10	0.25 (constant 15 mph)	0
Electric bicycle (1 rider)	300	25	0.3	20	20	0.4 (constant 20 mph)	15
Electric scooter (2 in tandem)	600	45	0.5	20	20	1.0 (typical driving)	50
Neighborhood electric vehicle (2 occupants)	7,500	600	11	25	30	1.9 (constant speed of 20 mph)	95
Smart car (2 occupants)	12,000	825	52	90	>300	16.0 (EPA urban cycle)	750
GM EV1 electric car (2 occupants)	600/ month	1,350	100	90	80	5.0 (EPA urban cycle)	250
Full-size SUV (7–8 occupants)	35,000	2,500	239	>100	>300	59.0 (EPA urban cycle)	2,727

Chapter 4

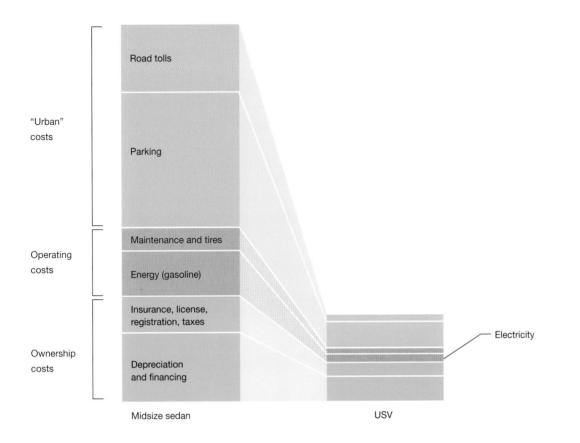

"Urban" costs

Operating costs

Ownership costs

Road tolls

Parking

Maintenance and tires

Energy (gasoline)

Insurance, license, registration, taxes

Depreciation and financing

Midsize sedan

USV

Electricity

costs when considering affordability, but there are many other ownership and operating costs that should be factored in as well. The American Automobile Association annually studies the various costs associated with owning and using an automobile[4] and concluded in 2008 that driving a 24 mpg midsize sedan for 15,000 miles per year incurs an average cost per mile of about 55 cents, assuming that gasoline costs $2.94 per gallon (i.e., 12.3 cents per mile). This equates to $20–25 per day. In other

Figure 4.21
Qualitative comparison of the total costs of owning and using a midsize sedan versus a USV. Total ownership costs in 2008 associated with a typical midsize sedan driving 15,000 miles per year with gasoline at $2.94 per gallon (costs associated with parking and tolls are based on New York City).

words, the fuel cost is less than a quarter of the total cost. (AAA's breakdown was shown earlier in figure 2.2.) Insurance costs, although difficult to predict, could also be substantially reduced if the likelihood of a collision decreases significantly.

Calling this the "total ownership cost" is not entirely accurate, as anyone who has parked a car in a densely populated city knows. According to the Colliers International 2008 Parking Rate Survey,[5] the *median* monthly cost for parking in Manhattan is around $500 (hence the $6,000 annual cost in figure 4.21). The cost is lower in other U.S. cities but can still be significant: $310 per month in Chicago, $350 per month in San Francisco, and $460 per month in Boston, for example. However, on the flip side, average monthly parking costs in other major cities around the world can be even higher than in New York: over $1,000 per month in London, and around $750 per month in Hong Kong and Sydney. On top of this, there are often charges to be paid for driving into big cities via bridges or tunnels, for example, and a few cities have even introduced congestion charging for driving inside the city itself. The toll charge for entering Manhattan each day is nearly $10, whereas the London Congestion Charge is $10–15 per day. For someone driving each weekday these additional charges can add up to roughly $2,500 per year.

The purpose of these calculations is to underline the affordability of USVs for urban use. Their high "fuel economy" (up to 200 mpg gasoline equivalent) and ability to use off-peak electricity could make them very affordable to operate, saving thousands of dollars a year in energy costs alone and perhaps recoup-

ing the vehicle purchase cost in two to three years. Moreover, in order to encourage their use, cities may exempt them from congestion charging. Even if they do not receive perks like free parking, one can expect their smaller footprint to translate into significantly reduced parking costs. For example, if five times as many of them can be parked in the same land area as conventional automobiles, the cost of parking should be approximately one-fifth and result in potential savings of several hundreds of dollars a month. These vehicles could also be given free access to dedicated lanes or HOT lanes. Insurance costs should be significantly less if they are involved in fewer collisions because of their crash-avoidance mechanisms.

A particularly important point to note is that, with electric USVs, the energy cost of operation (that is, the yearly electricity bill) is only a small percentage of a much-reduced total. This is a game changer. It not only reduces the burden of personal urban mobility on the world's energy supply, but also reduces the sensitivity of driver behavior to fluctuations in energy cost. With gasoline, as we have frequently seen, increases in price at the pump can motivate drivers to reduce their travel, they can reduce desirability and value of property in urban fringes that rely upon access by automobile, and they can have significant political repercussions. With USVs, drivers are more likely to be motivated by opportunities to reduce travel time and minimize unexpected delays—opportunities that are provided by wireless connectivity combined with the electronic management technologies that we shall introduce in subsequent chapters.

Summary: USVs Designed for Cities, Not Cities Designed around Cars

The conventional automobile is a marvel of utility, since it provides virtually unlimited ability to access the entire roadway system and can convey its occupants safely and comfortably over long distances at high speeds while carrying substantial amount of cargo. In a sense, automobiles are mostly driven around urban areas, but have been designed for intercity use. This versatility comes with a price in terms of cost, mass, size, and efficiency that is particularly acute in dense urban environments. A case can be made that cities, particularly those with reasonably wealthy populations and intense competition for limited land space, will increasingly shape the automobile's form and function just as the automobile shaped the city layout and landscape in the twentieth century.

Figure 4.22 summarizes the ways in which electrification, connectivity, and strategies for small vehicle design can combine to create an attractive alternative to the conventional automobile—the electric USV. Each one of these elements of reinvention can provide substantial benefits on its own, but in combination they add up to much more.

Electric-drive vehicles are clean, compact, and a pleasure to take out on the road. Their perceived high cost can be an issue when they are required to provide around 100 miles or more battery-electric range at highway speeds. But when much lower ranges and speeds are acceptable, as they can be for urban use, the cost of electric-drive vehicles can be

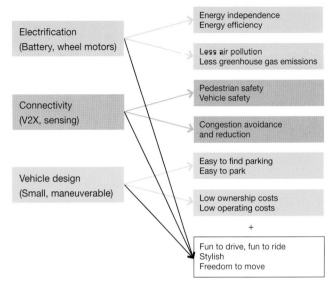

Figure 4.22
Benefits provided by the automobile's new DNA.

low, as exemplified by the extreme case of electric bikes. USVs are more expensive to own and operate than electric bikes, but much less expensive than traditional automobiles.

USVs, then, can be affordable, clean, efficient, maneuverable, easy to park, comfortable, safe to other road users, and a lot of fun. What's more, they can be expressive, fashionable, and desirable in a whole new way. Their combination of qualities, particularly well suited to urban environments, holds the potential for the fundamental transformation of personal urban mobility systems.

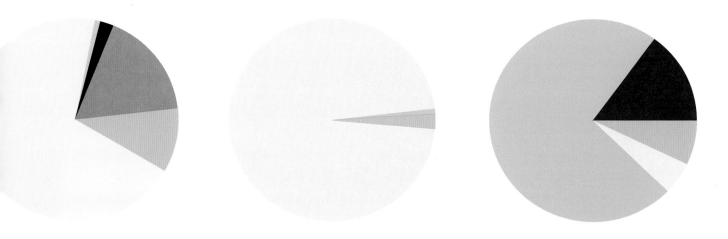

Clean, Smart Energy Supply

Automobiles need energy to power their wheels. It matters where this energy comes from, how it gets from its sources to the wheels, how much of it is wasted along the way, what side effects result, and what it ultimately costs both drivers directly and—in the form of externalities—society as a whole.

In this chapter we demonstrate the advantages of switching from gasoline to electricity for powering automobile wheels, and explore the practical issues that must be dealt with in doing so.

The Disadvantages of Gasoline

The cost of the energy required to power automobile wheels has long been a concern. One of the reasons for the rapid growth of automobile use from the early twentieth century onward—particularly in the United States—has been the low price of gasoline at filling stations. This has enabled the development of metropolitan areas with extensive low-density, automobile-dependent fringes. These, in turn, magnify the effect of any increases in energy prices, as recent periods of high gasoline prices have demonstrated. When energy prices rise, cities become more expensive to operate—with effects rippling throughout the economy: personal mobility is increasingly restricted; low-density urban fringes become less attractive; property values at urban edges tend to fall; and lower-income families who have sought relatively inexpensive housing in outer suburbs are disproportionately affected.[1]

A major disadvantage of powering wheels by burning gasoline in internal combustion engines, though, is that this process produces both local air pollution and carbon emissions that drive increases in greenhouse gases. Recent decades have seen considerable progress in reducing local air pollution from automobile tailpipes, but the problem of carbon emissions has become the focus of increasingly urgent concern. Schemes to tax, cap, and trade carbon have the long-term goal of reducing carbon emissions by creating incentives to reduce gasoline dependence, but in the short term they have the disadvantage of driving up gasoline prices and mobility costs.

Another well-known disadvantage of gasoline is that it is not renewable. The Earth has a finite stock of oil, and as this stock is consumed, the cost of discovering, extracting, and utilizing the remainder rises. There is controversy about how much petroleum remains for potential human use, what it will cost to make use of it, and whether global petroleum production has now peaked; but there can be no doubt about the long-term desirability of shifting to renewable energy sources. These sources—such as hydroelectricity, solar energy, wind energy, and biomass—are not diminished by consumption, and they are likely to become less rather than more expensive as the associated technologies improve.

Yet another disadvantage is that petroleum sources are concentrated in relatively few geographic locations, and this creates major energy security concerns for petroleum-importing nations. They become very vulnerable to interruptions in petroleum supply and price escalation. There is a fast-growing need, in many parts of the world, to achieve greater energy security through the utilization of more diverse and more widely distributed energy sources.

For all of these reasons, the petroleum-dependent global energy supply system that powers the wheels of today's cars is not sustainable. And the problems with it will only get worse over time. We propose to remedy this not only by replacing the internal combustion engines of automobiles with electric motors, but also by integrating electric-drive vehicles into new kinds of urban energy systems that are distributed rather than centralized, that make increasing use of diverse, clean, renewable energy sources, and that provide urban mobility at much lower cost than gasoline. These systems are enabled by the combination of smart automobiles that can buy, store, and sell electricity with smart-grid technology and dynamically priced electricity markets.

New Energy Supply Chains for Automobiles

To see how this is possible, we need to consider not only the power trains of automobiles, but also the supply chains that bring energy to them from primary sources. In principle, these chains might be very short, as with vehicles directly powered by sails or by solar panels on their roofs. In practice, though, they often extend back to distant primary sources and involve many storage, transfer, and conversion steps. Figure 5.1 shows the many ways in use today, or projected for the future, to power automobile wheels.

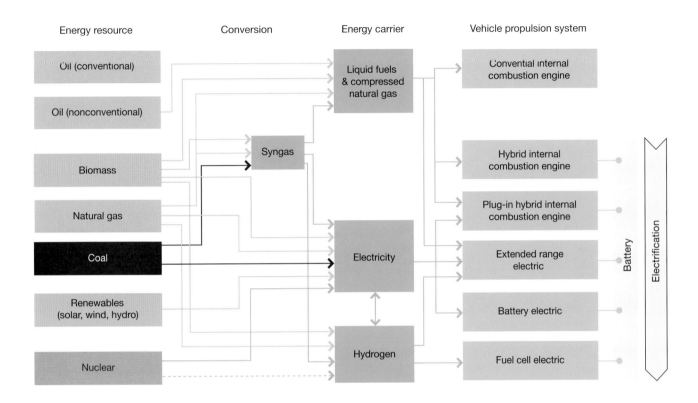

Figure 5.1
Supplying energy to power automobile wheels.

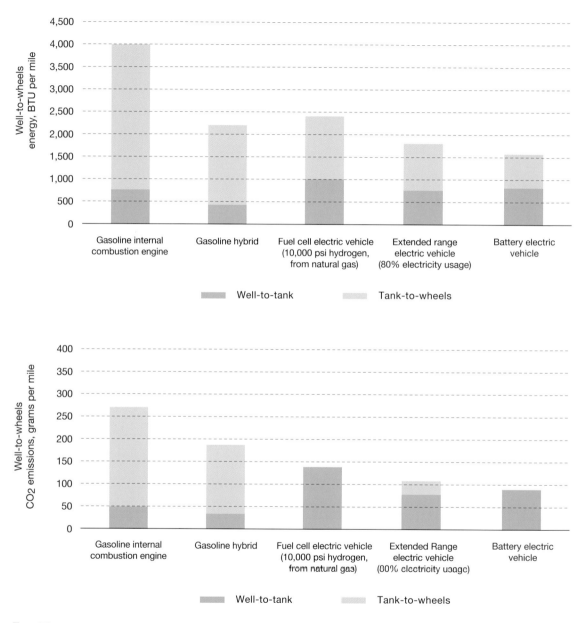

Figure 5.2
Tank-to-wheels and well-to-wheels energy
efficiencies and CO_2 emissions for various
propulsion systems.

Notice that there are, potentially, three major energy carriers in the system—liquid fuels (primarily gasoline), electricity, and hydrogen—the last of which, as we shall see, is closely related to electricity. Liquid fuels require an infrastructure of pipelines, refineries, and static and mobile storage reservoirs—culminating in the fuel tanks of automobiles. Hydrogen also requires the storage and transportation of liquid or gas. Electricity requires an infrastructure of generators, transmission and distribution cables, power and control electronics, and batteries or other onboard storage devices. Electrification—the conversion of energy from many sources into electricity and the distribution of energy in that form or as hydrogen—reduces or eliminates many of the problems with the sourcing, processing, distribution, and combustion of liquid fuels.

Each of the three approaches to electric vehicles that we have discussed (hydrogen fuel cell, battery, and extended-range electric) has a different vehicle efficiency (tank-to-wheels) as well as a different efficiency in getting the energy source to the vehicle (well-to-tank); the combined effect is called well-to-wheels efficiency (figure 5.2).

The Complementary Nature of Electricity and Hydrogen

The alternative to batteries, for providing electric power to a vehicle, is the combination of hydrogen storage with fuel cells. Batteries are charged directly using electricity, whereas fuel cells produce electricity when fed with hydrogen fuel (providing a "refillable" battery, in other words).

Like electricity, hydrogen can be produced from many different energy sources, and since hydrogen can be extracted from water using electricity (a process called electrolysis), any renewable pathway to electricity is also a renewable pathway to hydrogen. In this way, hydrogen and electricity can be viewed as interchangeable and complementary. Of course, there is an energy loss involved in this additional electrolysis process, but the advantage is that the vehicle can realize long-range (more than 200 miles) zero-emissions driving with a short refuel time (five minutes).

As battery and hydrogen fuel cell technologies advance, it becomes clearer that each technology is important to the success of the other. Each brings unique advantages to the vehicle—batteries offer lower operating cost in terms of energy used (but long recharge times and low energy density that limits vehicle range); fuel cells and hydrogen storage provide greater vehicle range and shorter fueling times (but require a new fueling infrastructure). Batteries work particularly well for small electric-drive vehicles such as USVs, but hydrogen fuel cells provide attractive trade-offs in larger vehicles such as family-sized vehicles and buses. Taken together, batteries and hydrogen fuel cells can optimize the use of diverse energy sources in support of varied transportation needs.

Hydrogen is a particularly valuable option because it can not only be made from water and electricity, but can also be derived from any hydrocarbon, such as biomass, natural gas, or coal. This ability to produce hydrogen from a broad portfolio of available energy sources makes it a very attractive component of efforts to effectively address energy diversity. The fact that water, electricity, and natural gas are widely distributed across the country also means that every home and business has access to all the raw materials needed to produce hydrogen.

A recent National Research Council report, findings of the Congressional Hydrogen and Fuel Cell Technical Advisory Committee, and collaborative research by General Motors and Shell have all concluded that a hydrogen infrastructure for automobiles is economically viable and technically feasible. In fact, for less money than was spent to build the Alaskan pipeline, conveniently located hydrogen stations could be deployed in the 100 largest U.S. cities and every 25 miles on all interstates, putting hydrogen conveniently in reach for 70 percent of the U.S. population.

Importantly, this infrastructure would complement the electric grid from an energy-diversity perspective because domestically supplied natural gas and biomass are excellent sources of hydrogen, and hydrogen is an excellent way to store electricity produced from renewable sources like the wind and sun until it is needed. Electricity generation, hydrogen storage, and fuel cells can be colocated, or they can be distributed throughout electric grids and vehicle fleets as engineering and cost considerations dictate.

The Advantages of Electrification

The most fundamental advantage of electrifying the energy supply chain—that is, converting energy from many sources into electricity and distributing it in that form or as hydrogen—is that it provides higher end-to-end efficiency than when using liquid fuels. Along the way from primary sources to wheels, less energy is wasted and fewer harmful by-products are released.

This translates directly into lower total energy costs for personal mobility. Today's batteries are inexpensive to recharge from the electric grid, costing about two cents per mile for electricity. This is one-third to one-sixth the cost of driving a comparable vehicle on gasoline, depending on whether you're paying $2 or $4 per gallon at the pump. Unlike gasoline tanks, which can be refilled indefinitely, though, batteries can only be recharged some finite number of times until they eventually need to be replaced. But even when you factor that in, electricity is still less costly than gasoline for powering automobiles right now, and its advantage will only grow over time.

A second advantage of electrification is that it enables energy diversity—the flexibility to use many different energy sources, all of which have their advantages and disadvantages, in many different locations. This allows long-term evolution of the supply system toward increasingly efficient, cleaner, more sustainable operation. It allows localities to take advantage of their particular energy opportunities in providing mobility rather than depending on importing oil from distant and often insecure sources. For example, West Virginia has abundant sources of coal, Hawaii mostly uses petroleum, Rhode Island uses natural gas, Idaho has vast hydroelectric resources, and Vermont draws on nuclear power for most of its electricity (figure 5.3). By integrating

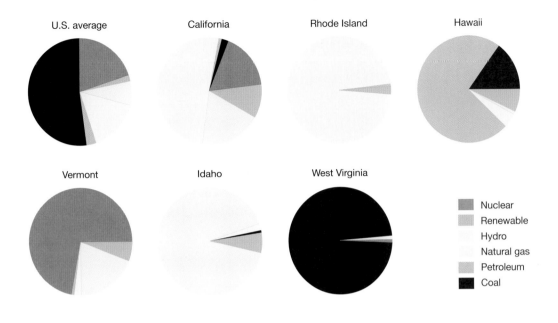

Figure 5.3
The primary sources of electricity for the U.S.
and various states.

many source types, locations, and supply routes, this diversity provides energy security and domestic employment.

A third advantage is that the onboard process of converting electrical energy into power to wheels is clean, silent, and highly efficient—in contrast with the combustion of gasoline, which, despite many decades of research and technological improvement, remains relatively noisy, dirty, and hot. This makes electric-drive vehicles a much more benign presence in urban areas.

And a final major advantage is that electrification simplifies automobiles as discussed in chapter 4: they have fewer parts, and far fewer moving parts.

The Effects of Energy Density

Battery-electric vehicles will not provide good performance, and will not be cost-effective, if their batteries are too bulky, heavy, and expensive (figure 5.4). Early battery-electric cars could not compete with their gasoline-powered rivals and quickly disappeared from

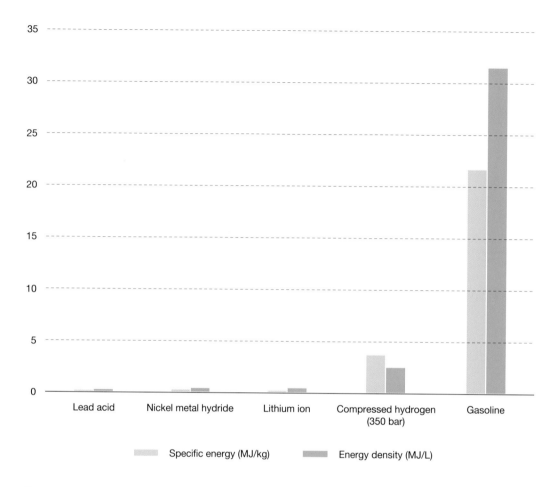

Figure 5.4
Energy density comparison of batteries, hydrogen,
and gasoline.

the marketplace, because their lead-acid batteries were not compact and light enough. They forced designers into an unattractive trade-off—either create vehicles that were overstuffed with heavy and expensive batteries or accept impractically limited ranges. Typically, however this trade-off is resolved, batteries end up occupying a far larger proportion of space within an automobile than gasoline tanks do (figure 5.5).

More precisely, since there are practical limits on the overall dimensions of automobiles, excessive battery bulk generally means that there is less space inside for passengers and baggage. Moving from internal combustion engines to more compact electric motors does open up more space for batteries, and shifting these motors out to the wheels helps even more; but reducing battery bulk to an acceptable level remains an important goal in the engineering and design of electric-drive automobiles.

Battery mass has a different effect. Battery packs can add hundreds of pounds to the mass of an automobile, and this additional mass must be accelerated—consuming more energy than would otherwise be necessary and requiring a more powerful propulsion system. Conversely, reducing the mass of an automobile's battery pack has the compounding effect of also reducing energy requirements and demands on the propulsion system. As described in chapter 4, when vehicle performance requirements are reduced then the energy required to accelerate and propel the vehicle's mass can be reduced, and this allows a smaller, less expensive battery pack to provide the required driving range.

Figure 5.5
The lithium-ion battery pack of the Chevrolet Volt (shown on the left) provides it with a range of 40 miles, roughly as much as one gallon of gasoline.

The Opportunity of Evolving Battery Technology

Battery technologists try to overcome these difficulties by finding ways to store electricity in as little space, and adding as little mass, as possible. As they achieve better energy–volume and energy–mass ratios they begin to open up new vehicle design opportunities and expand the market opportunities for battery-powered vehicles.

But this is not enough by itself. In addition to providing acceptable energy density, batteries for electric-drive automobiles must meet other practical requirements. They must be safe under reasonable operating conditions. They must be sufficiently durable—that is, provide a sufficient number of charge–discharge cycles before their performance degrades unacceptably. And, of course, their cost must be sufficiently low.

Figure 5.6 illustrates the evolution of battery technology and recent progress toward achieving these goals. Lead-acid batteries are inexpensive and are still widely used in some applications, for example in electric bikes in China, but they have many disadvantages. Nickel-metal hydride batteries are frequently used in today's hybrid cars, such as the Toyota Prius. These provide better energy density but at a higher cost. Lithium-ion batteries offer still higher performance and have been widely used in laptop computers and mobile phones.

In recent years, lithium-ion batteries have developed to the point where they can be considered for practical, large-scale automobile use, and they are likely to continue to improve. They now enable small, lightweight, battery-electric automobiles that combine sufficient space for passengers and baggage with sufficient range to be widely attractive.

Summary: Battery-Electric Automobiles Can Effectively Meet the Needs of Today's Urban Drivers

A combination of battery progress and a clear understanding of personal urban mobility needs (presented in chapter 4) leads to the conclusion that battery-electric USVs can, right now, meet the needs of urban drivers sufficiently well to provide an attractive alternative to today's gasoline-powered automobiles. The first generations of these vehicles will be able to perform well enough to initiate the switch to battery-electric automobiles, and their advantages will increase over time as further technological innovations, together with economies of scale, take effect.

The key is to combine lightweight vehicles, providing sufficient range for urban driving, with modest amounts of lithium-ion battery storage. This combination makes the vehicles described in chapter 4 highly practical. They can be inexpensive to purchase and operate, safe, convenient, fun to drive, energy efficient, and carbon free.

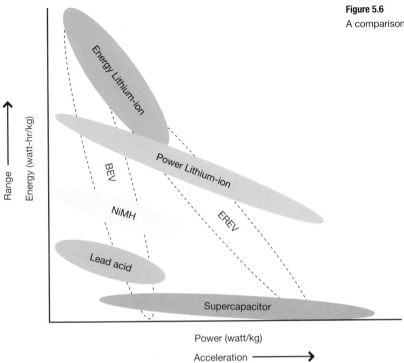

Figure 5.6
A comparison of automobile battery technologies.

Charging Infrastructure

Just as gasoline-powered cars required the creation of an extensive infrastructure for distributing gasoline and selling it at conveniently located filling stations (about 170,000 of them, currently, in the United States), and large-scale use of hydrogen fuel cell vehicles awaits the creation of hydrogen distribution and storage infrastructure, so battery-electric automobiles require an infrastructure of recharging locations.[1] This creates a chicken-and-egg situation: electric-drive vehicles require a charging infrastructure in order to become attractive and grow in numbers, while investment in a charging infrastructure must be justified by having electric-drive vehicles on the road in sufficient numbers.

Fortunately, in modern cities, the core of the necessary distribution system already exists in the form of the electric grid, which has been evolving for more than a century and is now ubiquitous.[2] It can, with some effort, be adapted and extended for the purpose of charging electric-drive automobiles. The key technical, economic, and design question is how best to transfer electricity from the grid to vehicle batteries.

The task is quite different from that of distributing and transferring gasoline (figure 6.1). For safety and to achieve economies of scale, gasoline must be stored in bulk at relatively few locations within a city, and these locations require tanker access. From there, specialized pumps transfer it to the gasoline

Gasoline	Electricity
Stored at few locations	Available at many locations
Fast transfer into vehicles	Slow transfer into vehicles
Unlimited number of refills	Finite number of recharges

Figure 6.1

Comparison of refilling with gasoline and recharging with electricity.

tanks of cars. By contrast, electrical outlets are ubiquitous, connections to them are much more easily made, they need only unobtrusive wires—not tanker access—and they make it economical to transfer small quantities of electricity. So we can have a much finer-grained, more distributed system.

There are also important differences in source-to-vehicle transfer rates. It takes only a few minutes to refill an automobile's gas tank at the pump. This means that trips can be interrupted for quick refills along the way, and that cars rarely have to queue for very long at service stations. But battery recharging can take much longer—often many hours. This is due to the limitations of battery chemistry and charging devices—the "pipes" that move electricity from the grid to the batteries.

Finally, refilling and recharging operations have different effects. A gasoline tank can be refilled or topped up an unlimited number of times. But a battery is designed to provide some finite number of recharges before it must be replaced and recycled. Furthermore, patterns of charging and discharging can have significant effects on battery life. This means that battery management is a significant issue for electric automobile owners and operators.

Design Requirements for a Charging Infrastructure

Several considerations intersect to determine how a city's recharging infrastructure should be designed and deployed. How far can electric-drive automobiles travel without running out of battery charge? If they have long ranges then it may be possible for the charging stations to be relatively sparsely spaced (although this will increase the time required to find a convenient station). If they are limited to shorter ranges then the stations must be more densely spaced. Whatever the spacing of stations, someone must find and pay for the necessary real estate and supply the grid connections and charging equipment, so policies on use of public and private space in cities, and the business models of equipment suppliers and charging station developers, will exert a strong influence on deployment patterns.

Where should the recharging take place? Logically, the charging infrastructure should first be installed where vehicles spend the majority of their time parked, and therefore have the greatest opportunity to acquire the energy they need. An analysis of National Household Travel Survey data indicates that the usual suspects—home, workplace, shopping

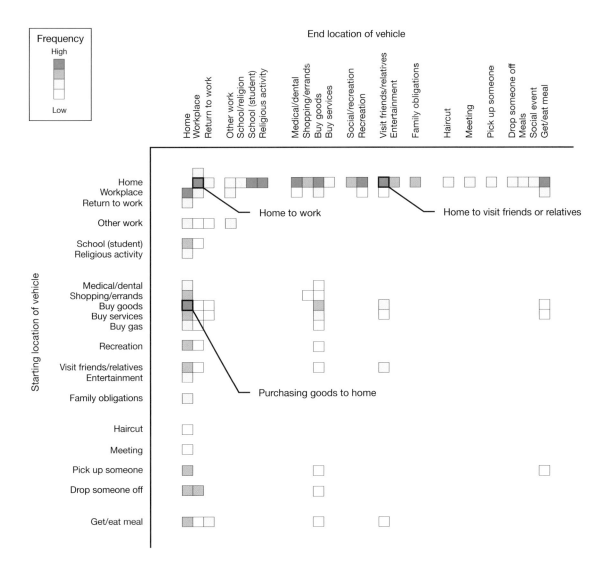

Figure 6.2
Locations where people leave their vehicles for at least thirty
minutes (source: McKinsey and Co. analysis of NHTS data).

malls—are the locations where charging capability would be most useful today in the majority of American cities. In the chart in figure 6.2, rows represent where trips originate, columns represent where they end, and the color represents the frequency with which the trip is made. For example, commuting from home to work is colored dark because this is a very common trip. Other common trips include visiting friends and going to or returning home from the store.

In interpreting these data, though, we should keep in mind that urban activity systems and land use patterns have always coevolved with mobility systems, and that this pattern is not inevitable and may not remain stable even where it currently prevails. In some contexts, for example, mixed-use, live-work "urban villages" and revitalized downtowns are challenging the "bedroom suburbs," with the result that daily commutes to distant workplaces become less important. Shopping centers with large parking lots exist primarily to serve the needs of automobile-dependent bedroom suburbs, and conditions for them may be less favorable in the future. In increasingly popular transit-oriented developments, automobiles may spend a lot of time parked at transit stops. Where shared-use automobiles are popular, vehicles spend a large proportion of their parked time at pickup and drop-off stations rather than homes.

In addition to location, speed of recharging is another consideration. Slow charging—overnight, for example—means that charging stations should be at locations where vehicles are parked for long periods, whereas quick charging means that recharging can be more casual, opportunistic, and on-the-fly.

How expensive is the charging equipment? When chargers are expensive, their cost must be justified by heavy use, so charging will tend to be centralized at popular locations. But when chargers are cheaper, they can be provided cost-effectively at more locations. In general, as determined by local circumstance, there will be some optimal mix.

Daily patterns of vehicle use also affect the recharging infrastructure strategy. Automobiles that are used for daily urban commuting can, with existing battery technology and acceptable battery pack sizes, be recharged overnight from standard 110-volt outlets to achieve the relatively modest running times they need each day—maybe a couple of hours to travel less than fifty miles to and from work. But other shared-use vehicles may run nearly continuously throughout sixteen-hour days, and may cover hundreds of miles in doing so. These need either much longer ranges between nightly recharges or frequent, fast recharges during the day. (There is a close analogy, here, with the electric forklifts used in factories and warehouses. Unlike private automobiles, these may work three shifts and thus need to minimize downtime for charging.)

Whatever the recharging strategy is, it should not be allowed to create "range anxiety" among drivers—the worry that they will run out of electricity and be stranded. The necessary sense of security can be provided either by automobiles with running times that are sufficiently long relative to intervals between opportunities for slow recharging, or by recharging points that are sufficiently densely grouped and offer quick recharging whenever drivers see that they are running low on charge.

Charging Infrastructure and the Grid

It is obvious that the design and deployment patterns of a city's charging infrastructure must meet the requirements of drivers in their daily lives. Less obviously, but equally importantly, the charging infrastructure must match the capabilities of electric grids.

First, there are issues of grid capacity. Fortunately, supplying 110-volt outlets for overnight charging should present no problems in most contexts. But use of higher voltages for faster charging may propagate consequences back to the substation level, and even beyond.

Second, there are issues of electricity trading and load balancing. Electric-drive automobiles can not only draw electricity from the grid, they can also store it, in some cases produce it onboard, and sell it. Making optimal use of these capabilities will be an increasingly important issue as electric-drive automobiles and cleaner, more sophisticated electric grids evolve together.

Following from this, the amount of time that an automobile spends connected to the electric grid, and the distribution of that time over the day, affect its capability to trade electricity—to buy, store, and sell it in advantageous ways as prices fluctuate. As we shall see later, this can be important in minimizing the cost of electricity to drivers, keeping the operation of the grid efficiently balanced, and making effective use of clean but intermittent energy sources such as wind turbines and solar panels. If an automobile is not actually connected to the grid when the sun is shining, for example, it cannot take advantage of the energy that this makes available.

Third, there are issues of electric supply quality. Will the introduction of large numbers of electric vehicles in charging stations end up degrading the quality of electric supply from the grid? Alternatively, can we take advantage of the storage capacity that they introduce to help regulate frequency and voltage, and help provide the kind of high-quality power supply that modern electronic devices and smart buildings require?

Electric utilities will clearly be powerful stakeholders in the development and deployment of the charging infrastructure, and strategies for this will need to engage their interests and concerns.

Slow versus Fast Charging

The patterns of initial deployment of the charging infrastructure, and of its long-term evolution, will be determined by shifting balances of the performance and cost trade-offs that we have described. The short-term implications of these trade-offs are reasonably clear, but there are many uncertainties in the longer-term outlook, and discussions of it must take account of these.

Today's commercial lithium-ion batteries (and alternatives to them), for example, are slow to charge—typically requiring eight or more hours to completely recharge an automobile battery pack from a 110-volt outlet. Furthermore, chargers that can reduce this time are expensive—typically about $1,000 extra for a 240-volt charger, for example. (One reason for this is that they require specialized power electronics—larger "pipes" to transfer electricity rapidly from outlets to batteries. Another is that the rapid transfer of large quantities of electricity creates danger that must be managed.) So the initial model for transferring electricity into electric-drive vehicles is likely to be slow charging at owners' homes overnight, perhaps augmented by some hours of daytime charging in workplace parking lots and garages.

However, there has been some exploration of deploying faster chargers. A partnership of Nissan and Ecotality (a supplier of chargers) has announced a plan to deploy electric automobiles and charging stations in Phoenix, for example. The idea is to offer the options of twelve-hour charging from 110-volt outlets in home garages, four-hour charging with home chargers that cost between $500 and $700 extra, and half-hour full charging or ten-minute top-off charging at public locations with $15,000 chargers.[3]

In Japan, plans have been announced to deploy hundreds of "quick recharge" stations. Tokyo Electric Power (Tepco) plans to supply $36,500 charging devices that can provide enough battery charge in a five-minute stop to drive a small electric-drive automobile for 40 kilometers, and enough in a ten-minute stop to drive 60 kilometers.[4] However, when the costs of additional grid capabilities to sup-

port these chargers are counted in, the total costs of installing them may be significantly higher. A five-minute recharge stop is comparable to that needed to refill a gasoline tank, and the cost of the charging devices may be roughly comparable to the costs of a filling station's underground tanks and pumps, so this suggests a pattern similar to that of today's filling stations.

Another possible alternative that has lately attracted some attention, and may have potential, is to exchange discharged batteries for charged batteries at swapping stations. This works well for laptop computers, electric bicycles, and scooters, where the battery packs are small and can easily be removed and replaced by hand. It is much more difficult with automobiles, which have larger, heavier battery packs and may not all be the same size and shape.[5] The battery-swapping approach imposes constraints on automobile design; it requires specialized mechanical devices to accomplish the swapping; it is very difficult to accomplish in the times necessary for "gas station" operation and is likely to produce queues of automobiles at swapping stations; and it introduces potential problems of mechanical reliability.

In general, the ideal of providing fast chargers for automobile batteries at many convenient locations does not present an insoluble technical problem. The problem is that the costs—for the chargers, their installation, and the additional grid infrastructure that may be necessary—are higher than for slower chargers, and in practice there may not be (without subsidy, at least) a sufficiently attractive business case for them.

Contact versus Inductive Charging

In a standard electric outlet, direct contact of conductive materials enables the flow of electricity. This is straightforward and efficient, and many of the electric vehicles in use or in development today use cables to plug into outlets that work in this way. There are ongoing efforts to establish standards for this.

Another possibility—as used in electric toothbrushes and their holders—is to transfer electricity through induction, without direct contact of conductive materials. Inductive transfer requires a primary coil on the grid side and a pickup coil on the vehicle side. It can take place across small air gaps, or through materials such as plastics. GM's early electric automobile, the EV1, employed inductive charging ports and paddles.

An advantage of inductive charging is that it opens up a wide variety of design options for placement of primary coils in parking places and pickup coils in vehicles (figure 6.3). Primary coils can, for example, be placed in the floor, and pickup coils in the undersides of vehicles. This eliminates the use of cables, plugs, and sockets, and enables charging simply by positioning a vehicle over a coil.

Another advantage is that primary and pickup coils can easily and reliably be sealed. This enables waterproofing, protection from vandalism, and operation under harsh and uncontrolled conditions.

Inductive charging devices, however, tend to be bulkier and more expensive than contact-charging equivalents. Their transfer rates tend to be lower. And, so far, there has been less effort to establish standards for their use with automobiles.

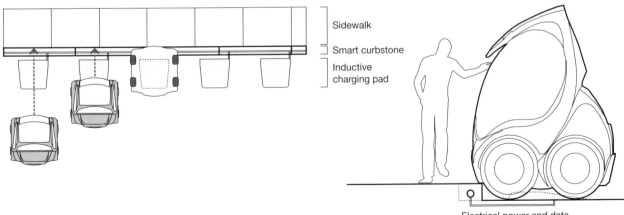

Sidewalk

Smart curbstone

Inductive charging pad

Electrical power and data

Figure 6.3
Inductive charging of a CityCar.

The Implications of Improving Battery Technology

It would be unwise to base strategies for initial deployment of a charging infrastructure on unrealistic assumptions about the capabilities and costs of batteries, power electronics, and electricity grids. There are not going to be any physics-defying miracles. But it would be equally unwise to assume that today's constraints will always apply. As with the growth of the Internet—aided by extraordinary advances in processing power, storage capacity, and bandwidth—the growth in electric-drive vehicles connected to smarter electric grids is likely to motivate large investments in research and development, and produce significant long-term performance improvements.

In particular, constraints imposed by battery chemistry may not be with us forever. Recently, for example, MIT researchers have developed a way of charging and discharging lithium-ion batteries at much higher rates.[6] This technology is still in its infancy, and it should be noted that fast-charging batteries do not suffice in themselves; they would have to be used in conjunction with fast chargers, as described earlier.

These sorts of technologies could not only shorten automobile charging times, they could also accept the high currents produced when electric motors act in reverse to provide regenerative braking—which requires the capability to absorb intense power bursts and would improve energy efficiency. In discharge mode, they could also enable the delivery of rapid power for race-car-like bursts of acceleration. A growing market for electric automobiles is likely to stimulate their development.

If recharging times and costs diminish, new possibilities for deploying charging infrastructure could begin to open up. In particular, dependence on recharging for long periods at homes and workplaces could become less of a factor. In the long term, electric-drive vehicles will probably be able to recharge frequently and automatically in parking spaces—either for full charge or top-off. The possibilities of keeping batteries continually topped up by charging whenever they are parked, and of opportunistic recharge just about anywhere, should significantly reduce range anxiety—the worry about getting stranded without the possibility of recharge.

Electrification of Parking Garages

Gasoline-powered cars have familiarized the ritual of filling up at gasoline stations. Electric-drive cars eliminate this, of course, but substitute different automobile and driver actions. These need to be designed carefully.

One simple way to initiate recharging is to plug in to a charger, as with any household electric device. This option has often been demonstrated. However, it requires the driver to think about it and perform a task—and, worse, creates a risk that the driver will forget to perform it. A more convenient and less risky option is to provide an automatic charging mechanism, as with warehouse robots that automatically plug themselves in. This is certainly feasible under controlled parking structure conditions, but it is more of a technical challenge for street parking spaces where the implications of water and snow, dogs, and vandalism must be considered.

Another possibility is to substitute inductive charging, as discussed earlier, for conductive charging, and simply park vehicles in close proximity to induction coils. Primary coils might be located in floors or streets to serve pickup coils on the undersides of vehicles (as with some inductively charged electric-drive buses), or in vertical structures to serve pickup coils in vehicle noses or tails.

Existing garages and carports can be retrofitted with charging devices, and the additional cost will often be sufficiently small, relative to the cost of the real estate and construction, to make this an attractive investment for property developers and garage operators. Where construction costs are of the order of $1,000 per square foot, and chargers cost somewhere around $1,000 per space, the cost increment for providing chargers is not large. And, where a parking space in a popular location returns somewhere around $100 per day in parking fees, the combination of higher occupancy rates and higher parking fees that electrified spaces can command can quickly translate into attractive profits. (These numbers are very approximate, and subject to significant change with circumstance, but you can plug in your own to test the robustness of this quick analysis.)

In "free" parking structures and lots, such as those in many shopping centers, the cost of providing parking for customers is part of the cost of doing business. Business owners will be motivated to invest in electrification when analysis shows that this will drive sufficiently many more customers to their doors—or prevent sufficiently many customers from being attracted away by competitors.

There are likely to be additional incentives, as well, to electrify parking garages. For example, the Leadership in Energy and Environmental Design (LEED) green building rating system, developed by the U.S. Green Building Council, has been a powerful motivator of green construction. LEED support for "green garages" would provide a strong stimulus to their development. Furthermore, municipalities that want to meet energy and carbon emissions targets are likely, at some point, to begin requiring green garages in their building and planning codes and enforcing these requirements in the permitting process.

The reduced footprint of electric USVs has the additional advantage of freeing up garage space for other uses. The real estate value released in this way can be very significant, and could also defray the cost of parking space electrification. In new dwellings, garages can be smaller, and because USVs are clean and silent, they can be integrated more closely with living space.

Enclosed garages provide opportunities for recharging under climate-controlled conditions and avoiding rain and snow. This can be particularly important in very hot and very cold locations, since battery charging typically does not work well at extreme temperatures. Under the same conditions, it also makes sense to precool or preheat the automobile's interior using the same battery-charging system, as this will reduce energy usage when driving and help to extend the driving range.

Smart Streets

Some urban areas rely almost entirely on off-street parking structures and lots, some rely almost entirely on street parking, and many have a mixture. In any case, strategies for the electrification of parking spaces must consider the two kinds of spaces working together as a system.

One difference between street parking and off-street parking is that street parking spaces are generally more exposed to the weather and to vandalism. Thus, on-street charging devices need to be more robust and carefully protected, and this drives up costs.

Another difference is that street parking spaces are usually publicly owned. Although they can provide significant revenue streams to municipal governments (in parking fines as well as parking fees), they respond primarily to the needs of residents and businesses that front streets. If there is insufficient street parking, residential property values and businesses tend to suffer.

Public ownership means that investment in electrification must be justified as a use of taxpayer dollars. It also means that home and business owners who depend on street parking (and who may have nowhere else nearby to get vehicles charged) have a direct stake in electrification plans. And it provides an opportunity for municipalities and lo-

cal political leaders (particularly mayors) to take the lead in promoting the use of clean, green, electric-drive vehicles.

Electrified street parking spaces would also become new elements in increasingly complex street infrastructure systems that serve multiple purposes. These now include street lighting and emergency alarm systems (both going back, in some locations, to the nineteenth century), traffic signals, street furniture and bus stops, electronic advertising, sensors, security cameras, shared vehicle stations, and wifi and cell-phone infrastructure. All of these require electric supply and communications, and they offer many opportunities to combine functions. As new streets are built, and as existing streets are upgraded, there is an opportunity to replace these tangles of overlaid, independent systems with integrated "smart streets" that provide all the necessary electric supply, sensing, and communication functions. Street charging stations are most appropriately viewed as elements of these new, integrated street infrastructures and potential catalysts for their creation.

In particular, there is an opportunity to combine the functions of parking meters and vehicle chargers. Since electronic parking meters have begun to replace mechanical models, both require electric supply. Both need to sense and meter automobile presence and consumption of resources. Both, when linked to telecommunication networks, can provide valuable, real-time data to managers of parking space and electric utilities. And there are obvious economies in combining two functions within one unit.

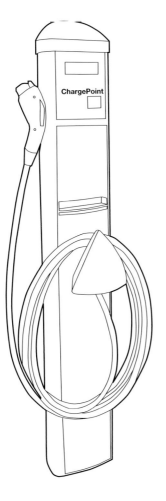

Figure 6.4

"Pillar" design for charging stations, as deployed by Coulomb Technologies in San Francisco.

Early in 2009, San Francisco initiated a pioneering, pilot deployment of public, street charging stations. The charging units, developed by Coulomb Technologies, were deployed in a two-year pilot by a partnership of the city, car-sharing fleet operators, and the electric utility company. The stations have a "pillar" design (figure 6.4), and electric-drive vehicles plug in to them. Their software is designed to meet the needs of drivers, utilities, municipalities, and parking space owners.

Smart Curbs

When electric-drive USVs provide front entry and exit, and are short enough to park nose-in to the curb within the width of a standard parking bay, they change the game. The underside of the car becomes the logical charging point, which opens up new design options.

Specifically, there is an opportunity to develop "smart curbs" that carry electrical supply and charging points, fit under the USV's nose, and connect to charging points on the underside (figure 6.5). Waterproofing is an obvious design challenge, but this does not seem insurmountable, particularly if inductive charging is used. Installation costs can be low, since installation can be scheduled for when streets are dug up and repaired. In the worst case, they only require removing and replacing existing curbstones—not entire paved areas. They have the enormous urban design advantage that they do not consume additional street space and do not obstruct pedestrian movement.

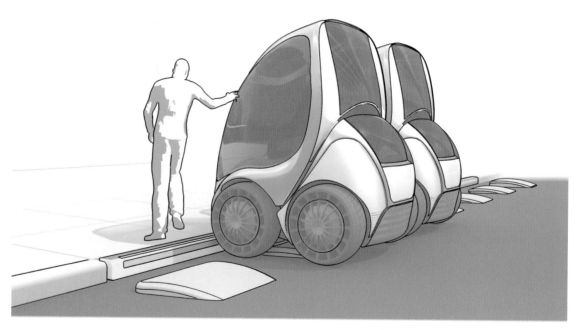

Figure 6.5
On-street parking with smart curbs for
inductive charging.

Figure 6.6
Inductive charging walkways can be overlaid on the slabs of existing parking structures.

In large-scale parking lots and parking structures (both existing and new), analogous "smart walkways" can be laid over the slabs (figure 6.6). Like smart curbs in streets, these carry electric supply and charging points, and they provide the additional advantage of separating pedestrian and vehicle traffic.

Potential Electrification of Roadways

In the long term, it might be possible to extend the electric supply infrastructure even further by providing charging strips or studs at appropriate locations in roadway surfaces. The most attractive locations for this would be high-investment, heavily traveled, tightly controlled roadways, such as the bridges and tunnels leading into Manhattan, and the extensive tunnels of Boston's Big Dig.

Technological feasibility has been demonstrated in some recent electric tramway systems, particularly in the APS (Alimentation par Sol) system used by trams in Bordeaux since 2003.[7] This system employs a rail in the road surface, divided electrically into eight-meter segments with three-meter neutral segments in between. For pedestrian safety, it senses when a tram is above a segment, switches it on, and then switches it off when the tram moves away from the segment. For transfer from these energized rail segments, trams employ collection skates mounted close to the road surface. This arrangement is unlikely to supply electricity at a fast enough rate, in the times that vehicles would be in contact with it, to provide full battery recharges. But it might extend range by reducing drain on batteries and providing top-off.

In 2009, KAIST (Korea Advanced Institute of Science and Technology) demonstrated prototype electric vehicles that were powered inductively from coil lines in the road surface (figure 6.7). However, it remains to be seen whether roadway charging

Figure 6.7
KAIST demonstration of electric vehicles being charged from an induction strip in the road surface.

systems can be made economical for many smaller vehicles, rather than relatively few large vehicles as with trams. Automobiles charged in this way require shorter charging segments than trams, and they are likely to be less precisely positioned. However, combination of charging strips with electronic guidance may resolve this difficulty.

Transfer from electrified roadways can be accomplished by means of some contact device (as with the Bordeaux trams) or through induction (as with the KAIST cars). An interesting long-term possibility (still at the early research stage, and far from practical application) is to make use of Witricity—a combination of induction and resonance developed by MIT researcher Marin Soljačić that allows efficient charging, through the air, over much longer distances than are possible with traditional inductive charging.[8]

Combining parking space and roadway recharging provides a long-term but potentially practical path to eliminating the traditional need to choose between the flexibility of battery-electric vehicles and the efficiency of continuously supplied vehicles such as electric trams and trains. By creating vehicles that pick up energy whenever it is available and rely on batteries when it is not, we may ultimately get the best of both worlds.

Incremental Deployment of Charging Infrastructure

A traditional problem with new mobility systems is the need to make large investments in infrastructure before any mobility benefits result. But it is not necessary to build vast quantities of recharging infrastructure before the changeover to battery-electric automobiles can begin. It is possible to pursue flexible, incremental changeover strategies. We can begin with vehicles that are relatively heavy on batteries and require long charging times, and then move gradually to lighter vehicles as charging infrastructure density and extent build up, and as battery technology improves.

Geographically, infrastructure construction should follow the rational and equitable principle of providing low-cost freedom of movement in proportion to population density (figure 6.8). It should provide complete coverage in dense urban areas. In less dense areas, it should provide it in reasonably close proximity to major network links—much as with piped water, sewage, and electric light. Battery-electric vehicles are not well suited for sparsely populated areas where driving range is high, where space is readily available for larger vehicles, and where there are fewer issues with air quality. This is where the hydrogen fuel cell or extended-range electric vehicle makes more sense.

There are many players in the game of deploying a charging infrastructure, and progress will depend on recognizing their varied constraints, aligning their incentives, and effectively coordinating their efforts. Battery and charger providers can accelerate the process, and open up possibilities, by continually improving performance and pushing down costs. (Intense competition will drive this.) Electric utilities can begin to exploit the synergies between electric-drive vehicles and smart grids. Real

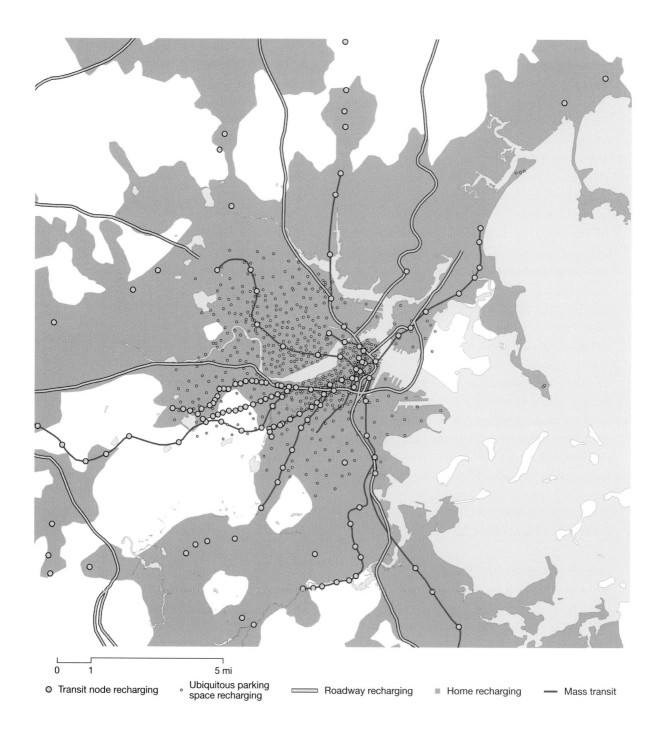

○ Transit node recharging	∘ Ubiquitous parking space recharging	══ Roadway recharging	▪ Home recharging	━ Mass transit

0   1   5 mi

estate developers and businesses that want to attract customers can pursue the competitive advantages and profit opportunities conferred by green garages. Regulators and standard setters can create incentives. Local governments can show leadership. And urban designers and infrastructure engineers can develop creative solutions to the complex problems of gracefully inserting this new infrastructure into urban space and operating it safely and reliably.

The process of deployment will be driven by a shifting balance of concerns. Drivers will want to minimize range anxiety. Electric utilities will want to take advantage of new opportunities to balance loads and improve the quality of electric supply. Merchants will want to attract customers. And governments concerned with sustainability will want to accelerate the uptake of electric vehicles by making them as convenient to use as possible in the early stages.

As with the development of the Internet, this process can be facilitated by open standards and positive network externalities. The more charging points that exist in the mobility network, the greater the benefit to you of investing in adding charging points at your own home or workplace. This, in turn, increases the value of existing charging points. And, as the network grows, electric-drive vehicles become lighter, more energy efficient, and more useful.

Summary: There Are No Insurmountable Barriers to Providing an Effective Charging Infrastructure

It is not necessary to invest massively in a charging infrastructure to initiate the large-scale use of electric-drive automobiles. Slow charging, from standard 110-volt outlets at homes and workplaces, will suffice in the beginning.

However, this charging strategy does have significant limitations, and as the use of electric-drive vehicles increases, investment in more sophisticated charging infrastructure will probably be justified. This will extend the charging infrastructure to parking spaces and structures, and eventually to roadways. It will probably reduce charging times—first to the hour or so that automobiles typically spend in urban parking spaces, and eventually to times that would allow opportunistic recharging and minimize vehicle downtime. And there will be a virtuous cycle of development: more electric-drive automobiles will create a demand for a more widespread and sophisticated charging infrastructure, while this infrastructure will increase the attractiveness of electric-drive automobiles.

Figure 6.8
Possible geographic distribution of charging infrastructure in Boston.

Integrating Vehicles and Smart Electric Grids

Historically, automobile-based mobility systems and electrical supply systems have been designed and operated as entirely separate entities. With the electrification of mobility, though, there are emerging advantages to designing and operating these systems in a carefully integrated fashion. Grid and power electronics experts, information technologists, and electric vehicle designers will need to work together to create a new type of urban system.

It is too limiting to think of electric-drive vehicles simply as consumers of electricity supplied by the grid. Parking large numbers of electric-drive vehicles in electrified parking spaces has the beneficial effect of introducing energy-storage capacity into the grid at a significant scale. Today's electric grids have little or no storage capacity, so this introduction helps with their well-known problem of "load leveling"—managing the troublesome dual condition of having idle generating capacity when demand for electricity is low and insufficient capacity when demand is high (figure 7.1). Furthermore, it does so very economically through dual functionality; vehicles serve as mobility devices when on the road and as energy storage devices when parked.

Load leveling is not the only potential benefit of tightly integrating electric-drive vehicles with electric grids. Grid experts point out that the availability of battery storage also facilitates voltage and frequency regulation. In general, electric-drive vehicles should not be treated as passive consumers of

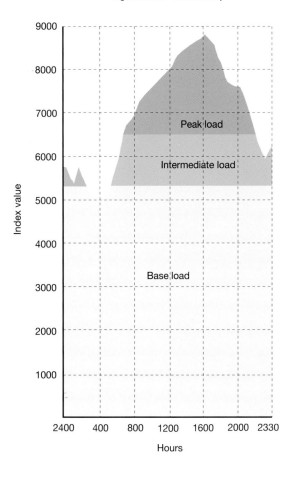

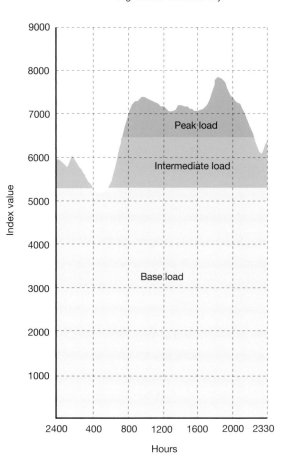

Figure 7.1
The load curves for a typical electric grid, showing the potential for leveling through electric-drive vehicles. Vehicles can buy and store electricity during load valleys and avoid buying or even sell during load peaks.

electricity, but should become active partners in the efficient management of electric grids.

Load-Leveling Strategies

The simplest and most obvious way to take advantage of the load-leveling capability provided by electric-drive vehicles is to recharge batteries in the early hours of the morning—when most vehicles are not in use, and electricity demand and prices are at their lowest. Vehicles could be programmed to recharge at particularly advantageous times; or, in more sophisticated systems, vehicles might monitor their patterns of daily usage and electrical price patterns, and be able to optimize their recharging patterns in response to these.

It is also possible, in principle, for electric-drive vehicles to sell electricity that they have bought from the grid at low prices back to the grid at higher prices at demand peaks. Electric-drive vehicles such as EREVs might also sell electricity that they have generated onboard. When you are away on vacation, for example, your temporarily stationary electric-drive vehicle might buy electricity during off-peak hours each day and then sell it back during peak hours. However, the cost to battery life of repeatedly charging and discharging must be counted against any gains from doing this. And the financial benefits from this sort of trading might not provide sufficient incentive to give up the flexibility provided by always having a fully charged battery pack. In the short term, then, selling back to the grid at peak demand times seems considerably less attrac-

tive than buying at off-peak times, but this may change as technologies develop and trade-offs shift. And electricity stored in car batteries can certainly provide emergency backup capacity.

These peak-shaving and valley-filling effects of electric-drive vehicles are not trivial, as some approximate calculations quickly show. A typical urban household might consume 10 kilowatt-hours (kWh) of electricity per day, and a vehicle's battery pack might store a comparable amount. The number of vehicles providing personal mobility in an urban area might roughly equal the number of households, so the vehicle fleet could store a significant fraction of the electricity to power all the households for a day.

The effectiveness of load balancing through electric vehicles will, however, depend a great deal on actual electricity prices. To illustrate this, consider the following example, based on 2009 U.S. prices. USVs will likely travel 8–10 miles on 1 kWh of electricity. The average cost of electricity is 8–10 cents per kWh. That means USV owners will pay a penny per mile for electricity if they use average priced electricity, and spend $100 for 10,000 miles per year. Now assume they always recharge off peak at 5 cents per kWh. They will save $30 to $50 per year with this practice (the cost of one to two tanks of gasoline). Now let's say that they can sell back 2 kWh during peak electricity demand each day at a price of 15 cents. They will pocket 20 cents from this transaction and give up 16–20 miles of range. In the absence of much higher electricity prices, or other incentives, this does not seem attractive.

This analysis, however, assumes individual ownership of vehicles, relatively infrequent recharging, and concern about range. In the context of large mobility-on-demand fleets, as discussed in chapter 8, optimizing electricity use will be important to fleet operators. And, in the context of ubiquitous automatic recharging, as discussed in chapter 6, range sacrifice may not be a concern.

There are many complexities still to be worked out. But it seems clear that, at a large-scale energy systems level, electric-drive vehicles can create a new and very significant advantage. When they are in motion, they are providing mobility. And when they are at rest and connected to the electric grid, they are providing the grid with much-needed storage capacity. Unlike today's cars, which spend most of their time unproductively sitting around, they never need to be idle.

Making More Efficient Use of Fossil Fuels

In practice, where will grids get the energy to power large numbers of electric-drive vehicles? The answer to this question determines whether electric-drive vehicles will yield significant reductions in carbon emissions.

In most contexts today, most of the electricity flowing into the grid is generated by burning fossil fuels—coal, oil, and natural gas—in large generation plants. This process also accounts for most of the world's carbon emissions. This is particularly a problem in China, where electricity production is very heavily dependent on coal. It is easier to control emissions at coal-burning generation plants than at automobile tailpipes, and it may eventually be possible to sequester the carbon produced at these plants; but for now, from a carbon-emissions perspective, substituting electricity from coal for gasoline just shifts a significant part of the problem from tailpipes to smokestacks.

We should, however, keep in mind the overall high energy efficiency of USVs. Even when coal is used to produce electricity for them, they will still produce far less greenhouse gas emissions than using gasoline to fuel traditional automobiles or even hybrid electric vehicles.

Energy experts have urged movement away from this long-established generation system, and they have suggested aggressive 2050 goals for doing so—for example, eliminating all fossil fuel deliveries to cities by that point. But there is large investment in existing plant and supply lines, and it takes time to develop and deploy adequate alternatives, so the changeover will inevitably be gradual. Realistically, we need incremental strategies for phasing out fossil fuels while phasing in alternatives as quickly as possible.

A good first step is to begin relocating combustion from moving vehicles to immobile installations within buildings. Combined heat and power (CHP) systems—both large-scale and small-scale—are already quite widely and successfully used. These employ the well-known principle of cogeneration for efficient, simultaneous generation of electricity and production of heat for hot water and space heating (figure 7.2).[1] In some versions they also

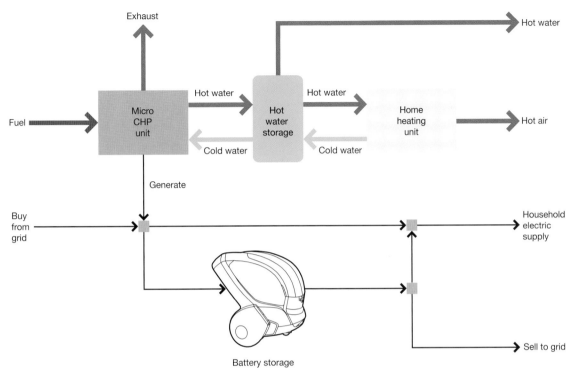

Figure 7.2

The principle of cogeneration applied in a domestic-scale solar heating and cooling system.

provide space cooling. They can be installed in individual buildings, or they can operate at neighborhood scales. The electricity that they produce can be used both to power buildings and to recharge the vehicles parked in their garages.

At the level of the conversion process, this is fundamentally more efficient than combustion of gasoline in millions of car engines, since it puts more of the heat generated by combustion to good use. Furthermore, by consolidating the combustion plants at far fewer locations (although more than with large, centralized generation plants), it enables economies of scale. By immobilizing the plants it saves the energy that would otherwise be used to move them around, loosens space and mass constraints on plant design, and simplifies fuel supply and storage systems—which need to reach fewer immobile locations rather than more mobile vehicles. And it enables more effective management of carbon emissions and other waste products.

The earliest cogeneration plants were large scale—serving university campuses and the like—and served primarily to generate electricity, with useful heat (for hot water and space heating) as a by-product. More recently, micro-CHP systems have become increasingly cost-effective. These are designed to operate in homes and small commercial buildings, and they reverse the emphasis—serving primarily to generate space heat and hot water, with electricity as a by-product. Typically, they generate more electricity than required to operate the buildings that house them, and this excess electricity can be used to charge electric vehicles in their garages.

This shift of combustion back into buildings represents a historic turning point. The process of combustion—together with the processes of fuel supply and combustion product removal—has been an integral part of urban life ever since early dwellings acquired hearths and chimneys. It migrated to vehicles in the early nineteenth century, when the first steam locomotives emerged. Now, after two centuries on wheels, it is time to repatriate it to its origins in buildings, where it is more easily managed.

Effectively Integrating Renewables

Progressively shifting to clean, renewable sources of electricity is an even more attractive option than relocating combustion. Over the entire energy supply chain for vehicles, this reduces fossil fuel consumption and carbon emissions to zero.

Cities and their hinterlands exist within fields of solar radiation, moving streams of water and air,

thermal gradients that extend down from the surface of the Earth, and sometimes geothermal energy. And they stick interception devices into these fields—roofs, towers, dams, wells, mines, and basements, in particular. It is obviously possible, then, to deploy solar collectors, hydroelectric and tidal barrage plants, wind turbines, heat pumps, and geothermal "heat mines." Why isn't this already happening on a larger scale?

The amount of energy incident on cities from these renewable sources is impressive, but the difficulty lies in collecting it. The efficiencies of available collection devices are limited, particularly under urban conditions that produce shadowing on solar collectors and turbulence that is bad for wind turbines. Thus the amount of electricity that can be generated per square mile from these sources is constrained, and its cost tends to be high. Still, as technologies improve, locally harvested, clean, renewable energy will have an increasing role to play in powering urban mobility.

An additional problem with solar panels and wind turbines is that they are only intermittently active. The sun does not necessarily shine when you want to drive your car, and the wind does not obligingly blow. Perhaps we could, like sailors, wait for appropriate travel conditions; but this does not seem feasible in modern urban contexts.

The usual practical effect of this intermittency is to add the cost of operating reserve (generators that are not in operation, but ready to be brought online as needed to meet peak loads), which is necessary to assure a reliable electric supply, to the already high

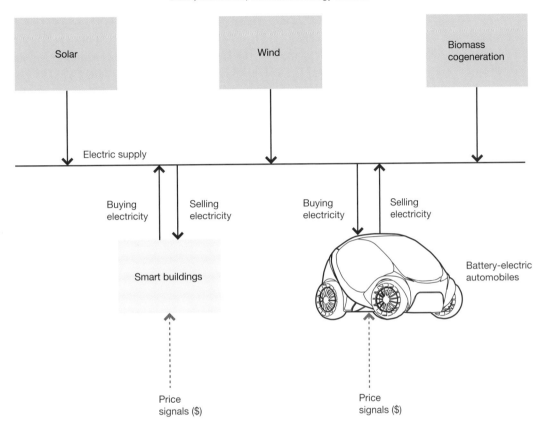

Clean, distributed, intermittent energy sources

Solar

Wind

Biomass
cogeneration

Electric supply

Buying
electricity

Selling
electricity

Buying
electricity

Selling
electricity

Battery-electric
automobiles

Smart buildings

Price
signals ($)

Price
signals ($)

Figure 7.3
The efficient combination of intermittent renewables
with battery storage in electric vehicles.

direct costs. But, in combination with the battery storage and load-leveling capability of electric urban mobility systems, solar and wind power can become much more cost-effective (figure 7.3). Solar panels, which usually provide maximum output in the early afternoon, can charge batteries for the evening rush hour. And wind turbines, which typically provide maximum output at night, can charge batteries for the morning rush hour. In general, intermittency matters much less when the grid incorporates cost-effective electricity storage at sufficient scale.

This, then, yields another important benefit from the close integration of electric-drive vehicles with electric grids. Electric-drive vehicles can not only make use of clean, renewable energy instead of fossil fuels, but can also—by providing battery storage "for free" (since the vehicles need batteries anyway) and thereby mitigating the effects of supply intermittency—enhance the effectiveness of grids that draw on renewable sources.

Distributed Urban Energy Systems

Another potential problem for large-scale electric mobility systems is that electric grids experience significant transmission losses when they transfer energy over long distances. (This is a particular case of the general rule that moving energy from where it is produced to where it is needed carries a penalty.) So it is difficult, for example, to efficiently power urban vehicles from hydro sources in distant mountains or from solar collectors in distant deserts. New transmission technologies may eventually reduce the magnitude of this problem, but they seem unlikely to eliminate it.

However, distributed systems of combined heat and power plants in building basements, solar panels on roofs, and wind turbines in suitable locations minimize transmission losses by distributing and mingling generation and consumption points within the urban fabric. Collections of these small, networked installations can serve, under unified control, as virtual power plants.

Integration of electricity storage further enhances the efficiency and reliability of such distributed systems. As we have seen, the batteries of electric-drive vehicles can provide part of this storage capacity. When these batteries have been used for a while, their performance may diminish to the point where they can no longer meet the demanding specifications for powering wheels, but they can still usefully be moved into building basements to serve as storage there. And excess electricity in buildings can be converted to hydrogen for storage, and then converted back to electricity in fuel cells—located either in buildings, or in fuel-cell vehicles that fill up from storage points in buildings.

A system of this sort is much more like the Internet than it is like an old-fashioned, centralized electricity generation, transmission, and distribution system or mainframe computer system (figure 7.4). It has devices that serve, store, and consume electricity. Some of these devices are immobile and others are mobile. And links among devices serve for two-way exchange as necessary, rather than one-way distribution from a production point to consumption points.

Figure 7.4
An Internet-like distributed urban energy system.

Eventually, these Internet-like energy systems can be integrated with similarly structured water and waste-recycling systems as well. Pumping water consumes fuel or electricity; moving water to elevated storage or heating it in insulated tanks stores energy; and running water through turbines (maybe in the reduction valves of urban water supply systems) can generate electricity. Urban waste can be recycled for burning in combined heat and power systems, or it can be used to produce methane, which is then used as fuel.

This all adds up, over time, to a revolution in basic urban services. Where water, gas and oil, electric, and waste-removal systems have traditionally been specialized in their purposes, highly centralized, and separately constructed and managed, they can now begin to converge and evolve into distributed, integrated, general systems for providing and distributing the energy that supports urban life. Potentially, these more integrated urban support systems are not only highly efficient, but can also achieve great robustness through their diversity and redundancy. Like the Internet, they can continue to operate effectively when parts of them fail or are destroyed, and when some of their supply streams are interrupted.

Dynamic Electricity Pricing

Obviously, the behavior of distributed urban energy systems, integrated with electric-drive vehicle systems, is complex and presents significant management challenges. On the supply side, the outputs of different kinds of generators in different locations fluctuate, and stocks of electricity or hydrogen stored at various locations go up and down. On the demand side, the energy requirements of buildings and mobility systems vary in daily and weekly patterns. The management problem is to keep supply and demand in equilibrium, so that consumers are never left without electricity to meet their needs and producers do not have to invest in excess capacity to guarantee this.

A useful management tool for this purpose is *dynamic electricity pricing* (figure 7.5). The idea is that electric utilities adjust, at closely spaced time intervals, the prices at which they sell electricity to consumers and buy electricity from small producers such as rooftop solar panels on houses. Prices go up when load on the grid is high, and they go down when demand is low. The management goal of this price adjustment is to keep load fluctuations within an acceptable band.

The underlying assumption is that there is considerable elasticity in consumer behavior. In response to price signals, consumers can decide—within broad limits—when to recharge their vehicle's batteries, when to operate their appliances (such as dishwashers), when to turn their air conditioning up or down, and when to sell electricity they may have stored in batteries or in a hydrogen fuel cell system. The optimal consumer strategy, obviously, is to try to buy when prices are low and sell when they are high.

Operation of this type of electricity market requires sophisticated information and communica-

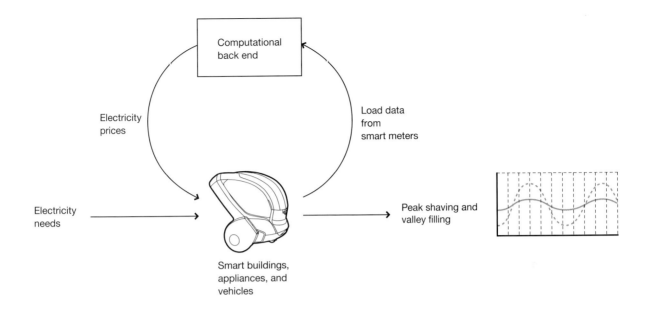

Figure 7.5
The use of dynamic electricity pricing to balance supply and demand.

tions technology. Electricity use, at numerous widely distributed locations, must be metered closely and reported back to grid operators in real time. Operators must process a massive amount of this load information, in real time, to calculate and adjust prices. And vehicles, appliances, air conditioning systems, and the like must be smart enough to execute optimal electricity-trading strategies—that is, they must be programmed to buy, store, and sell electricity, in the most advantageous way possible, in response to fluctuating and sometimes unpredictable price signals. The result is a real-time feedback loop, connecting electricity buyers and sellers, that acts to maintain equilibrium.

The Parallel Emergence of Electric-Drive Vehicles and Smart Grids

Unfortunately, in most cities today, electric grids are not sophisticated enough to support distributed generation and storage of electricity, buildings and vehicles that sell as well as consume electricity, and dynamically priced electricity trading by intelligent vehicles and buildings. For historic reasons, they are run by highly regulated monopolies that have the mission of assuring a ubiquitous, reliable supply, keeping electricity prices stable, and minimizing the risk of investing in new capacity. They use very simple electric meters to measure and bill for consumption, usually at fixed rates or rates that vary only in simple ways. However, active efforts are currently underway to develop and implement "smart grids" (figure 7.6) that provide the necessary capabilities (as well as other benefits that we do not need to go into here).

One crucial step, in the development of smart grids, is to replace old-fashioned meters with modern, digital meters that can provide frequent readings—every fifteen minutes, say—together with instant two-way communication between customers and the electric utility. This enables frequent adjustment of prices, together with billing based on this more dynamic price variation, and feedback to customers that enables them to optimize their electricity use in response to price signals.

Today's Grids	Smart Grids
Unsophisticated meters	Smart meters
Inflexible pricing	Dynamic pricing
Centralized generation only	Enable distributed generation
No storage	Incorporate storage
Unfriendly to clean, renewable, but intermittent energy sources	Friendlier to wind and solar energy
Minimal use of information technology	Utilize digital networking and large-scale, sophisticated computational back end

Figure 7.6
Comparison of today's electric grids and emerging smart grids.

Another step is to enable two-way electricity flow—not only to consumers from the grid, but also from solar panels on building roofs, local wind turbines, hydrogen storage sites, battery packs of vehicles, and so on, back to the grid. These flows back to the grid must also be metered at closely spaced intervals, and the selling prices to the grid must be adjusted.

A third step is to provide electric vehicles, air conditioning systems of buildings, home appliances, distributed generators and storage units, and other electrical devices and systems with the onboard intelligence to respond to price signals. Where they have the flexibility to do so, they should buy electricity when the price from the grid is low and avoid consumption or sell electricity when the price is high.

All this requires the creation of a high-speed electric system communications network and defines a massive data-processing and real-time control task (figure 7.7). The state of a very large, distributed, dynamic system must be sensed, a huge data stream must be analyzed in real time, prices must be computed, and price signals must be sent to spatially distributed sites of production, storage, and consumption. Grid operators must compute optimal prices, and intelligent devices must communicate with the grid and compute optimal responses. This is challenging, but it is increasingly feasible.

Smart grids can also provide tools for buildings and electric-drive vehicle owners to manage their energy consumption and carbon footprints. One way to do this is through Web portals that provide analysis and control tools. Instead of going to a gas station to refill a traditional car, for example, a smart-grid customer might go to a Web portal to specify a vehicle's battery recharging strategy based on price and the environmental impacts of a specified mix of electricity sources.

Currently, there are pilot implementations of smart grids in operation. Many companies are jumping into the development and production of smart-grid hardware and software, and it seems likely that implementations will soon scale up. Southern California Edison, Pacific Gas & Electric, and American Electric Power have all deployed smart-grid technologies. One of the most ambitious and interesting pilot implementations has been SmartGridCity in Boulder, Colorado. This is a project of Xcel Energy, NREL (National Renewable Energy Laboratory), GridPoint, and other partners, with a goal of integrating 50,000 residential, business, and light industrial customers into the system.

During the twentieth century, industrialized nations built two kinds of massive but disconnected energy conversion systems—gasoline-powered light vehicle fleets and electric grids. But that situation is about to change. There is, now, an emerging convergence of electric-drive vehicle and smart-grid technologies. They are maturing within the same time frame; each will be beneficial to the efficient operation of the other; each will facilitate the large-scale deployment of the other; and they are likely to be increasingly closely integrated with one another.

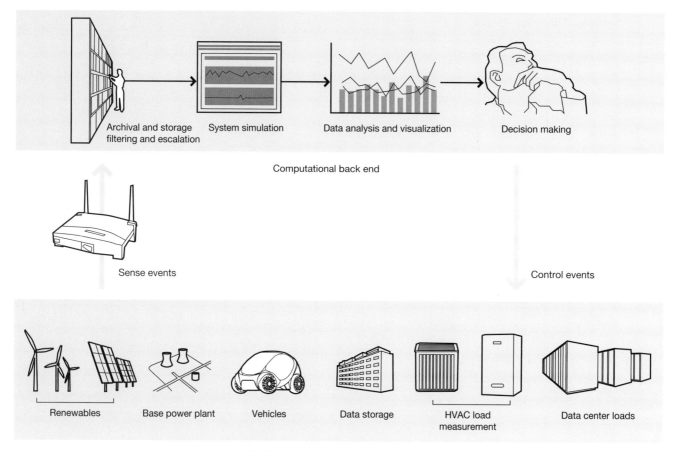

Archival and storage filtering and escalation

System simulation

Data analysis and visualization

Decision making

Computational back end

Sense events

Control events

Renewables

Base power plant

Vehicles

Data storage

HVAC load measurement

Data center loads

Electricity producers and consumers

Figure 7.7
The operation of a smart grid: production, distribution, and consumption under real-time control.

Chapter 7

Summary: Smart Sustainability

We can no longer afford to treat automobiles as disconnected mechanical devices that occasionally purchase bulk refills of energy that arrives through poorly structured, unstable, and vulnerable supply chains, at not particularly rational prices. Nor, if we want overall urban energy efficiency, can we simply focus on the efficiency of vehicles in converting energy supply into power to wheels—the last step in a very long chain.

Instead, we must begin to integrate the historically separate support systems of cities—fuel, water, and air supply; energy conversion; electrical and building services; and mobility—and bring them under unified control. We must create electricity markets that can respond effectively to needs that are unevenly distributed in space and vary dynamically in time, and that keep supply and demand in equilibrium through dynamic pricing and real-time feedback loops. And we must create vehicles that are alert, intelligent buyers and sellers within these markets.

Through shifting as much functionality as possible from moving vehicles to the fixed infrastructure of an integrated mobility and electrical system, vehicles can become simpler and lighter. They don't have to carry around a lot of batteries that add to vehicle cost, bulk, and weight, diminish energy efficiency, and eventually have to be recycled.

This is a strategy of systemic *smart sustainability*.[2] It seeks major efficiency gains by focusing on *overall* system performance, not only the individual performances of the various components and subsystems. And it uses ubiquitous digital networking and distributed intelligence—forming urban nervous systems—for the necessary level of fine-grained, highly responsive, real-time control of urban energy systems.

It is a large, long-term project, but it is not a utopian one. Investments in infrastructure renewal can begin to move us down the path toward its realization.

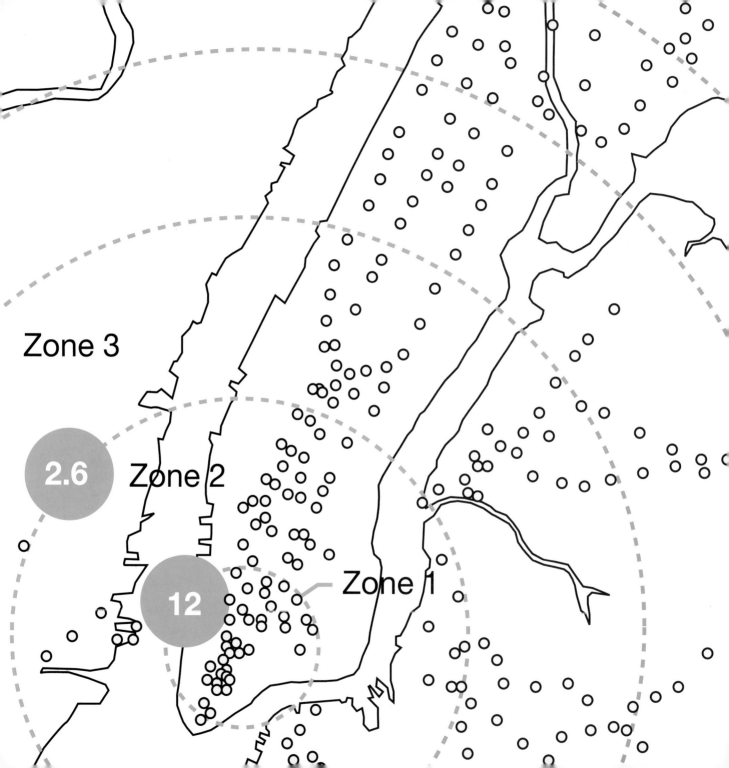

New Mobility Markets

Designing clean, safe, fast, fun, attractive, and inexpensive vehicles is one-half of the task of reinventing personal urban mobility. Integrating these vehicles into efficiently operated urban systems is the other half. After all, an electric-drive vehicle, even if the electricity comes from renewable sources, still gets the equivalent of zero miles per gallon if it is running its air conditioner while stuck in traffic.

In the last chapter, we saw how to make energy supply systems more efficient. Electricity supply and demand fluctuate; operators of smart grids can vary electricity prices with these fluctuations; and intelligent automobiles can respond to price signals by buying, storing, and selling electricity in what they calculate to be the most advantageous ways.

This creates a real-time feedback loop that keeps supply and demand in equilibrium.

In this chapter, we show how the same principle of dynamic pricing, combined with intelligent vehicles that enable drivers to respond appropriately to pricing, can be applied to the other basic resources required by personal urban mobility systems—road space, parking space, vehicle fleets, and insurance.

This broad application of dynamic pricing provides many advantages. Clear, rational, responsive pricing of trips provides a sound basis for both individual decision making and the optimization of overall system behavior for society as a whole. From a driver's perspective, it makes the total costs of trips accurately and clearly evident and enables well-informed choices among alternative trip departure times, routes, and destinations. From an urban systems perspective, it enables the effective management by price of available urban space and infrastructure while providing tools for achieving social equity and other policy objectives. From a business perspective, it opens up new opportunities to attract customers through context-sensitive advertising and price incentives. The overall result is a self-organizing personal urban mobility system that responds to varied needs while effectively minimizing demands on energy supply systems, urban space, vehicle fleets, and driver time.

We begin by considering road and parking space markets. Ideally, these should keep supply and demand in reasonable balance, and in doing so they should efficiently allocate available capacity to meet demands that vary, in complex ways, in space and time. They are reciprocally coupled: when cars are on the road they are not parked, and vice versa.

Drivers have a limited capacity to process information, so we cannot expect them to keep close track of varying road and parking prices and adjust their behavior accordingly. But, as we shall see, intelligent vehicles can make the task manageable.

Smoothing Peaks in Road and Parking Space Demand

The highly uneven distribution of demand for urban road and parking space, both spatially and temporally, has traditionally made it very difficult to allocate these resources efficiently.

Automobiles naturally converge on popular destinations within cities, creating road congestion in the vicinity of these while leaving other roads nearly empty. People want to park near these destinations, which has the effect of saturating nearby parking while leaving spaces elsewhere vacant. People want to travel during peak hours, and are less inclined to do so at other times. If roads and parking areas are sized for peak demands then they are uneconomically underutilized at other times; but if they are sized for average demands then they cannot accommodate the peaks. Figure 8.1 shows this effect, and illustrates the significant resulting increase in the cumulative real estate required to support parking, since each vehicle needs on average more than two parking spaces.

As with electricity supply, it helps to smooth out demand peaks. We can achieve this by taking advantage of redundancy in road and parking systems and elasticity in travel behavior.

Unlike highways that are designed for high throughput, urban street and road systems are typically highly redundant, providing many alternative routes to most destinations. Of course it is desirable to take the shortest route, but you might be induced to take a slightly longer route if it is significantly cheaper. This provides an opportunity to employ price incentives to achieve more evenly distributed, efficient use of available road space.

It is the same with parking. There are many parking spaces distributed throughout cities, some in close proximity to particular destinations and some more distant. You will be happiest with a

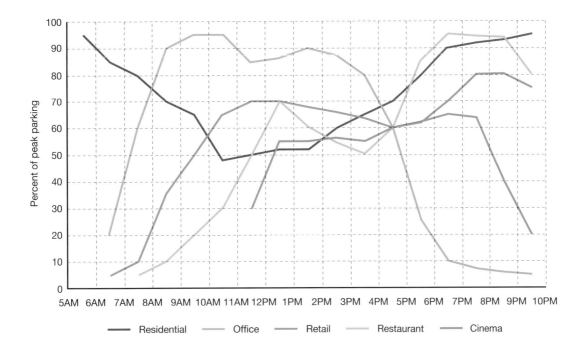

100
90
80
70
60
Percent of peak parking
50
40
30
20
10
0

5AM 6AM 7AM 8AM 9AM 10AM 11AM 12PM 1PM 2PM 3PM 4PM 5PM 6PM 7PM 8PM 9PM 10PM

—— Residential   —— Office   —— Retail   —— Restaurant   —— Cinema

parking space nearby your destination, but you might be induced to accept a more distant one if it is sufficiently less expensive.

Furthermore, there is some elasticity in mobility demand. Depending on trip prices, city dwellers may shift the times of their trips, opt for alternative destinations, or decide not to make trips at all. These are familiar phenomena in air travel, where demand management through pricing has long been commonplace.

Figure 8.1
Variation in parking demand throughout the day and by location.

Extending the Principle of Congestion Pricing

A first step in creating efficient, dynamically priced road and parking space systems is to track when and where cars use roads, and to charge tolls accordingly. This can be accomplished with GPS location-tracking technology combined with wireless connection to transmit road-usage data for billing. GPS truck tracking is now commonplace for the purpose of truck fleet management, and a GPS-based truck toll system has been in use in Germany since the mid-2000s.

Under standard implementations of GPS tracking, the location privacy of drivers is violated, and this is likely to be socially unacceptable. (The same is true of more familiar traffic monitoring technologies, such as surveillance cameras and toll transponders.) However, it is possible to encrypt time-position data in such a fashion that useful functions of a vehicle's path (such as speed and toll charges) can be computed without revealing anything more than the results of these computations.[1]

The next step is to adjust road prices according to congestion. When roads are heavily congested the price per mile of using them goes up, and when they are less congested the price goes down. The idea of congestion pricing as a means of managing road demand is well known and has been widely studied. The first large-scale implementation of congestion charging was introduced in Singapore in 1975. Perhaps the best-known example is London's Congestion Charge Zone, which started in 2003 and affects the central part of London (figure 8.2).

Existing congestion pricing systems are relatively crude, however. They only adjust prices at fairly infrequent intervals, and over fairly long road and street segments. Furthermore, they cover only parts of urban road systems, so they tend to displace traffic to other parts of the system. The idea here is to make congestion pricing systems citywide, fine-grained in their spatial resolution, and frequent in their adjustment of prices as congestion levels fluctuate.

This requires close, citywide monitoring of traffic flows. Such monitoring is already approached in existing urban traffic management systems, such as the Los Angeles Department of Transportation's ATSAC (Automated Traffic Surveillance and Control) system. It can be accomplished through an infrastructure of roadway induction loops (which has proven not to be very satisfactory), or through a combination of booms over roadways and transponders in vehicles, as in many automated tolling systems. And, as we have noted, it can also be accomplished through GPS tracking of vehicles.

Congestion may not be the only determinant of price. Planners might discourage travel through quiet residential areas, for example, by imposing road price premiums there. Conversely, they might encourage travel through other areas by providing discounts. Or, with electronic driver identification and suitable privacy protections in place, they might provide road price discounts (much as with discount cards on public transit systems) to the elderly, low-income service workers, people who ride-share, and so on.

Under this sort of system, then, the cost of a road trip to a motorist is the distance actually traveled

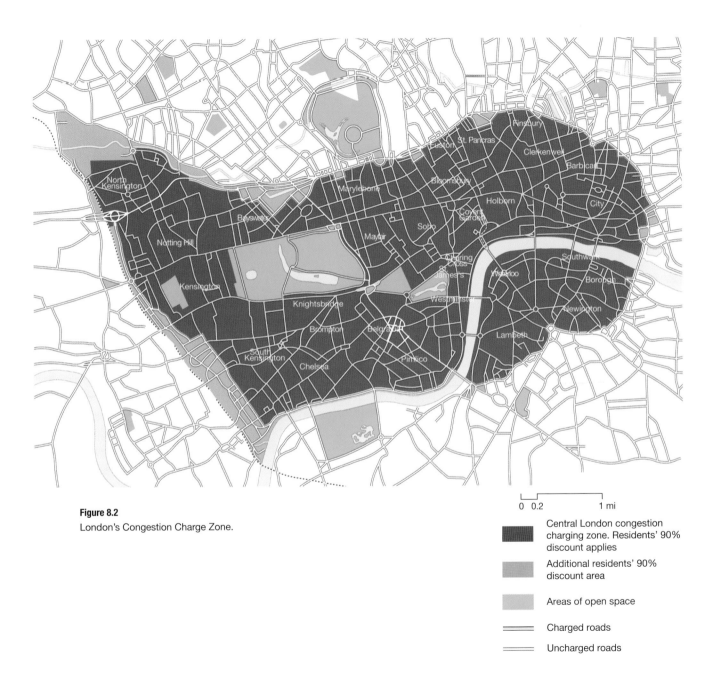

Figure 8.2
London's Congestion Charge Zone.

Central London congestion charging zone. Residents' 90% discount applies

Additional residents' 90% discount area

Areas of open space

Charged roads

Uncharged roads

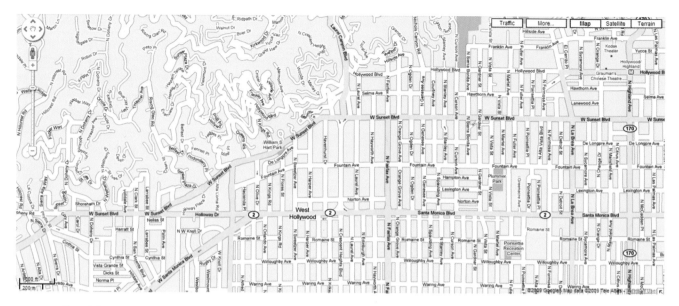

Figure 8.3
Street grids (at right) provide many possible routes to
destinations, but suburban cul-de-sac systems (at left)
often provide only a few, or even just one.

multiplied by the costs per mile—determined by congestion and other factors—of the road segments that are traversed. The primary effect is to create incentives for drivers to minimize travel through congested areas and at peak times, and to even out demands on available road capacity.

There are several ways for drivers to respond to these incentives. Sometimes they have the flexibility to depart earlier or later in order to take advantage of better prices. Sometimes they can choose differ-ent destinations—as when the mission of a trip is to buy a quart of milk at a convenience store, there are several stores within reasonable range, and the prices for getting to them vary. And, often, there are alternative routes to a chosen destination. This depends on the structures of street and road net-works; street grids provide many alternative routes, but suburban cul-de-sac systems, which have tree-like structures, may provide only one way to get to any particular house (figure 8.3).

Integrating Road Pricing, Navigation, and Optimization

For this sort of system to work, drivers need accurate, real-time information about trip options, prices, and times. Providing this information is a task for a large-scale, distributed computing and communications system.

When real-time road price information is available it can be broadcast wirelessly through such a system to the GPS navigation systems of automobiles. This allows the calculation not only of shortest-path routes to desired destinations, but also of lowest-cost routes. A driver might request the lowest-cost route subject to a travel time constraint, or the fastest route subject to a cost constraint, or the route that provides the highest likelihood of arriving exactly at a specified time. This creates a real-time, electronically managed feedback loop that regulates demand for road space (figure 8.4). The routes chosen by vehicles respond to the overall pricing in the system, while simultaneously, the overall pricing responds to the distribution of vehicles.

There is a complication, however. The relevant consideration is not the actual price of some distant route segment at the outset of a trip, but the expected price when the vehicle actually reaches that segment. So software for computing trip times and costs must take account of predicted congestion levels and prices—and such predictions will always have some degree of uncertainty.

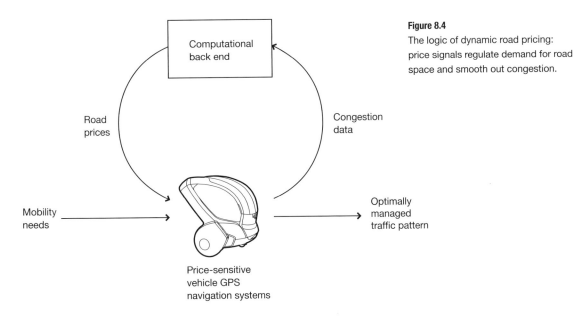

Road prices

Congestion data

Computational back end

Mobility needs

Price-sensitive vehicle GPS navigation systems

Optimally managed traffic pattern

Figure 8.4
The logic of dynamic road pricing: price signals regulate demand for road space and smooth out congestion.

A further complication is that drivers will often want travel time and cost certainty. So, in order to be really useful, routing algorithms need to compute paths that minimize the chances of unexpected delays and provide high likelihoods of getting you there exactly when they say they will. This is a difficult, nonlinear stochastic optimization problem, but there are algorithms for computing good solutions to it, and intelligent vehicles should be able to provide sufficient computation capacity to execute them. (A complementary approach is to employ electronic traffic management and crash-avoidance technology to eliminate events that cause unexpected delays.)

Road-space markets that are structured in this way contrast vividly with the markets that exist today, in which motorists either do not pay according to level of usage, or do so through fixed tolls and very simple congestion pricing schemes. Within more sophisticated markets, navigation algorithms can be tuned to the needs of individual drivers by, for example, minimizing the risk of unexpected delays for some and facilitating the search for bargains for others. They can evolve in sophistication, much like stock-trading algorithms. And the managers of urban road systems can establish pricing policies that create incentives for drivers to use the road system as efficiently as possible.

Extending the Principle of Dynamic Pricing to Parking

Today's urban parking space markets are even more poorly structured than road space markets, and can similarly benefit from dynamic, electronically managed pricing. In most cities today, the prices of parking spaces are either not adjusted in response to varying demand, or are only adjusted in accordance with very simple and inflexible strategies. In addition, drivers get very little information about space availability and price. Frequently, they find it necessary to drive around randomly until they discover available spaces, and this process is severely constrained by visibility limits. In dense urban areas, this parking search process adds significantly to traffic congestion, adds to the driving time required to reach desired destinations, and wastes energy and creates air pollution.

Imagine a more intelligently regulated system in which the occupancy of parking spaces by vehicles is monitored electronically, and information on available spaces is communicated wirelessly to GPS navigation systems. All parking spaces are dynamically priced, and there are electronically managed, eBay-style auctions on them. If you are desperate for a parking space, for example, you might instruct your vehicle to bid high—probably yielding you a space exactly when and where you want it. Conversely, if you are looking for a bargain, you can instruct your vehicle to bid low, and you will probably end up with a less convenient but cheaper space.

In combination with electronic identification of drivers, this system also allows implementation of socially or economically desirable parking policies. For example, residents of an area, or low-income service workers, might be provided with monthly allocations of electronic "parking chips" with which to bid. And merchants might provide chips to their customers.

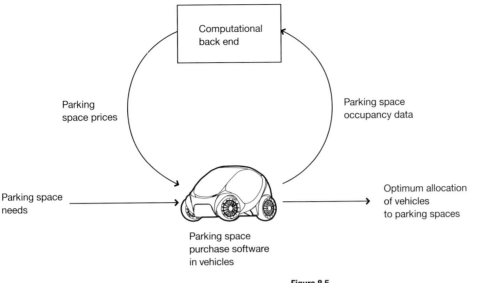

Figure 8.5
The logic of dynamic parking space pricing: price signals regulate demand for parking space and smooth out overcrowding.

As with electronic management of road space, this establishes a system with a real-time feedback loop (figure 8.5). Parking pricing responds to the demand expressed by individual parking decisions, while simultaneously, individual parking decisions respond to pricing.

Enabling Mobility-on-Demand Systems

Yet another component of overall trip cost—in addition to electricity, road, and parking cost—is the cost of the vehicle. When vehicles are privately owned, this cost is normally borne by the owner-driver. Furthermore, vehicle purchase and insurance are sunk costs even before a mile is driven. The alternative, for those who want access to automobiles but cannot or do not want to pay these costs in full, is shared vehicle use.

In principle, shared-use automobiles can provide a good match of capacity to need. For many urban dwellers, one automobile is too much but no automobile is too little to meet their mobility needs. And, even when they own automobiles, they do not always have convenient access to them—for example, when they are in cities far from home. Traditional sorts of shared cars—taxis, rental cars, distributed-access short-term rental cars as offered by ZipCar, and so on—are responses to these conditions. However, they are generally less convenient to access than privately owned automobiles. But intelligent, connected vehicles transform this picture. It would be possible to drive down the cost of vehicle sharing, efficiently match vehicle supply to demand, and provide very good mobility service by creating sophisticated, electronically managed mobility-on-demand systems. Within these systems, vehicle fleets, distributed at convenient locations throughout urban areas, are shared by large numbers of drivers.

The effects of this on the convenience, efficiency, and cost of urban mobility can be dramatic. There are always constraints on the numbers of vehicles that can be accommodated in urban areas and the numbers of parking spaces that can be provided. (These constraints are particularly severe where urban population densities are very high, as for example in Singapore.) Fractional, rather than integer, vehicle possession makes better use of limited vehicle and parking space stocks. Hundred-percent-owned vehicles spend most of the day (usually about 80 to 90 percent of it) parked, since their owners want to spend most of their time doing something other than driving. But fractional-possession automobiles can have much higher utilization rates—that is, they can spend more time on the road and less time parked. Instead of one person owning an automobile 100 percent of the time but only actively using it for 20 percent, five people might—with cleverly managed allocation—each use it for 20 percent.

When automobiles are in use on the road they are not occupying parking spaces, so higher utilization rates also mean lower parking space demands.

Or, to put it another way, more vehicles can be accommodated by a given number of parking spaces. This also means that all parking spaces are short-term spaces. Higher utilization rates result in higher turnover of parking spaces, with the resulting higher probability of finding a vacant space when and where you need it. Furthermore, when you want to pick up a vehicle to drive, you don't have to pick up your own particular vehicle—which may be at a distant location, requiring you to walk or have a valet retrieve it. The nearest available vehicle will do. This reduces door-to-door travel time, it is particularly convenient in poor weather, and it saves energy as well.

Finally, well-managed vehicle sharing provides a politically viable solution to the problem that dense urban areas currently attract more vehicles than they can comfortably manage. Simply letting vehicles pour in results in mobility system breakdown, which is self-defeating. Limiting parking is highly unpopular with merchants and with those who want convenient access to an area. Limiting automobile ownership through prohibitions or high taxes creates haves and have-nots. Congestion rings, like that employed in London, can be unpopular and are often politically impossible. But fractional vehicle access allows efficient and equitable distribution of a finite resource—the city's capacity to accommodate a personal mobility fleet.

Mobility-on-demand systems don't need to and shouldn't push these efficiencies to the point of creating rigidities and inequities. They should gracefully accommodate those who want to use automobiles as storage for a while—on multistop shopping trips, for example—and those who really do want to own an automobile 100 percent and are prepared to pay for it.

Successful mobility-on-demand on a large scale requires the support of several technologies and management strategies. These combine, as described below, to create mobility-on-demand systems that effectively and efficiently meet personal urban mobility needs.

Automating Pickup and Drop-Off Transactions

In traditional car rental systems, the pickup and drop-off transactions are cumbersome and slow. This creates a barrier to their use. So the first step in creating a mobility-on-demand system is to combine registration or membership systems with electronic identification and electronic tracking and billing to minimize transaction times and costs.

The model for this is one-click online purchasing, enabled by previous storage of your credit card number, your address, and other relevant data. In a mobility-on-demand system, you simply walk up to a conveniently located, available vehicle, swipe your credit card or otherwise electronically identify yourself to unlock it, and then take possession and drive away. At drop-off, the system automatically recognizes that you have relinquished possession and locks the vehicle again.

This requires sophisticated, behind-the-scenes information processing. But much of the necessary technology is already highly developed for use in cus-tomer identification, authorization, and credit card transactions, and it can be adapted for this purpose.

Creating and Managing One-Way, Distributed Rental Systems

A further impediment to use of traditional car rental systems is their requirement to return cars to where they were picked up. They are *two-way* rental systems. This simplifies the task of the system operator, but it is often inconvenient for users. Ideally, mobility-on-demand systems should provide *one-way rental*: just pick up a vehicle, drive it to your destination, drop it off, and forget about it. If you want a vehicle at some later point, just pick up another one.

Bicycle sharing systems, such as Vélib in Paris and Bixi in Montreal, have successfully pioneered one-way rental of personal mobility devices in urban areas. There have also been a few attempts to provide one-way rental services with traditional, gasoline-powered automobiles. In particular, car2go has introduced one-way automobile rental, with Daimler's Smart cars, in Ulm, Germany, and Austin, Texas. But electric USVs, which are extremely economical and can be recharged in their parking spaces, promise to be particularly suitable for one-way rental.

In one-way rental systems, users don't have to worry about retrieving and returning vehicles. Higher vehicle-utilization rates become possible since the vehicle becomes available for use by another customer as soon as it is dropped off. This reduces the time vehicles spend sitting around unavailable and not being used.

In addition, mobility-on-demand systems should provide access points conveniently distributed throughout their service areas—by contrast with most current car rental systems, which seek economies of scale by providing access at a few large, centralized locations with many vehicles. This minimizes door-to-door travel times, measured as the total time taken to walk to an available vehicle, pick it up, drive it to a parking space near the destination, drop it off, and walk to the destination. And if vehicles are able to drive autonomously, then they can be summoned from parking spaces at pick up and sent by themselves to parking spaces at drop off. This potentially reduces door-to-door times even further.

Retail location theory suggests that, where pickup and drop-off points are of equal capacity, they should serve equal population catchments. This means that they will be closely spaced in areas of high population density and more sparsely spaced in areas of low population density. Alternatively, pickup and drop-off points might be spaced evenly, at intervals determined by comfortable walking distance, and varied in size (numbers of vehicles and parking spaces) according to surrounding population density.

Another way to look at this is in terms of the time cost of walking (in dollars per hour), and the total cost of driving a mobility-on-demand fleet vehicle (time cost plus rental). The total cost of a trip is that of walking to a pickup point, driving to a drop-off point near the desired destination, and then walking from the drop-off point to the desti-

nation. Pickup and drop-off points can be located to minimize total trip costs for customers.

Often, existing features of the urban fabric will provide convenient locations for pickup and drop-off points. Figure 8.6, for example, illustrates a study of Taipei in which possible locations for shared electric vehicles (bicycles, scooters, or automobiles) are shown at convenience stores and bus stops. Convenience stores are a ubiquitous feature of Taipei, and there is a mutual business benefit in using them as access points for a mobility-on-demand system: the stores are at convenient locations and often have available parking space that can be repurposed, and they benefit when mobility access points bring customers to their doors. Locating access points at bus stops enables driving a mobility-on-demand vehicle to a bus stop, taking the bus for the long portion of a trip, and then picking up another mobility-on-demand vehicle at the other end.

If access points are sufficiently densely spaced, mobility-on-demand systems can provide better door-to-door times than privately owned vehicles. Available rental vehicles will often be closer than you could park your own vehicle. And any available vehicle will do—a particular advantage when heavy demand for parking spaces makes it likely that your own vehicle will be distant.

What if the vehicle you go to pick up is dirty, vandalized, or in need of repair? Wouldn't this possibility discourage you from using a mobility-on-demand system? There are several effective remedies. First, there will usually be more than one vehicle available at a pickup point, allowing you to report

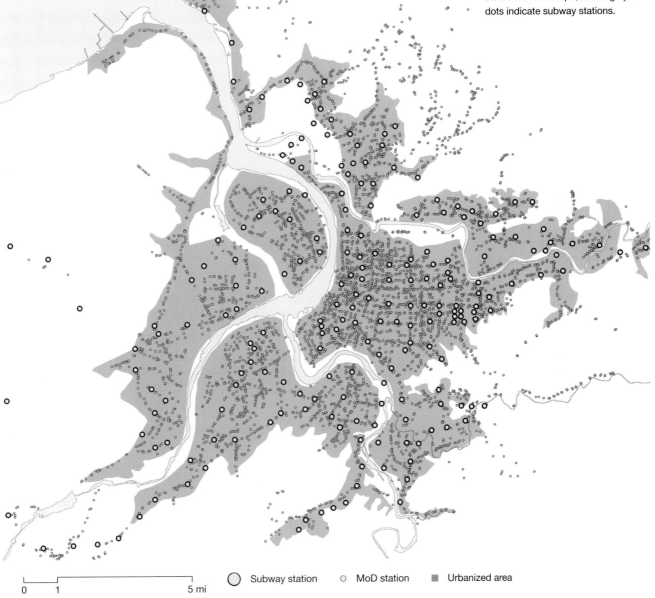

0	1		5 mi

◯ Subway station   ◦ MoD station   ■ Urbanized area

an unacceptable vehicle to the system operator and pick up another one. Second, the electronic tracking of vehicles means that the system operator always knows who has a vehicle at any moment and can determine responsibility for damage. (This also protects against theft.) Finally, within communities, social networks and peer pressure can often act to keep shared vehicles in good condition.

Where Mobility-on-Demand Systems Make the Most Sense

Appropriate densities and distributions of pickup and drop-off points also depend on the demographic and physical character of the urban fabric. In high-density, mixed-use urban areas, for example, origins and destinations are fairly randomly distributed and trips can occur anytime. This means that pickup and drop-off points should be distributed in approximately even patterns throughout service areas. This is the model generally used in bicycle sharing schemes that serve urban centers.

In areas where trains or buses provide efficient transit between stops, mobility-on-demand systems can efficiently provide "first mile" and "last mile" service. With pickup and drop-off points located at transit stops, they can fill the inevitable gaps between actual trip origins and destinations and the locations of these stops. This is particularly effective at the outer edges of transit networks, where the density of stops diminishes (figure 8.7).

In commuter suburbs, mobility-on-demand systems can connect transit stops to homes. On work-day evenings, commuters pick up cars at transit stops, take them home, and keep them overnight to recharge. Then, in the mornings, commuters can take cars back to transit stops. If the vehicles used for these short distances are USVs, then more commuters can access the transit system as more parking spaces become available at transit stations because of the USV's small footprint.

At a larger scale, where high-speed trains provide efficient intercity travel, mobility-on-demand access points at train stations can solve the first-mile and last-mile problems for train travelers. Figure 8.8 illustrates this concept applied to Taiwan's high-speed rail system.

Yet another, very simple strategy for implementing mobility-on-demand systems is to locate vehicle access points at dense concentrations of residential space, such as university dormitories and apartment towers. Residents can then pick up vehicles whenever they need them to run errands or attend events, and then return them when they come home. This sort of system treats the residential unit as a "base" for accessing the resources of the surrounding neighborhood, but provides much greater range from that base, and capacity to carry goods such as groceries, than is available to pedestrians. It may be particularly attractive to real estate developers. In 2009, for example, Audi Japan and Sumitomo Realty & Development initiated such a system for a residential and office complex in Tokyo's Roppongi district. Conditions seemed appropriate: parking costs are extremely high in Roppongi, and there were many short-term residents, who did not own cars, in the complex.

Subway stations per square mile in New York City

Zone	Subway stations	Radius (miles)	Area (sq. miles)	Stations/sq. mile
1	37	1	3.1	12
2	66	3	25.1	2.6
3	102	5	53.4	1.9
4	103	7	100.5	1.0
5	92	9	153.9	0.6

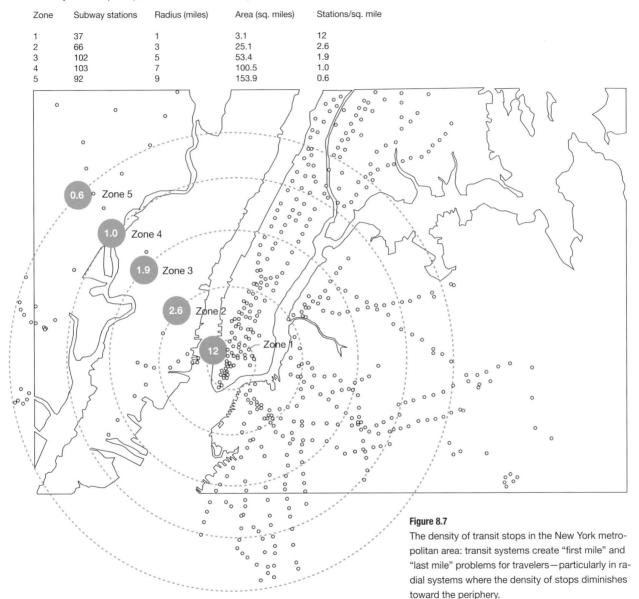

Figure 8.7
The density of transit stops in the New York metropolitan area: transit systems create "first mile" and "last mile" problems for travelers—particularly in radial systems where the density of stops diminishes toward the periphery.

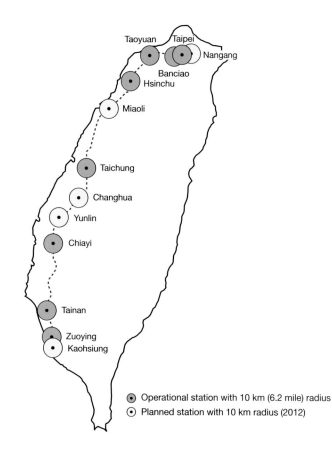

Figure 8.8
A study of a possible mobility-on-demand system to supplement Taiwan's high-speed rail line. Circles indicate coverage provided by mobility-on-demand access points at stations.

Rental rates can vary to manage overall demand for vehicle possession within these systems. To encourage short possessions, rates can be very low for, say, the first half hour, and then escalate. To smooth out demand, rates can be high during peak travel hours and lower during off-peak hours. To encourage use of the system by suburban commuters, rates can be very low during the late night and early morning hours—when, in any case, few people will want to take possession of cars and make trips.

Managing Vehicle and Parking Space Stocks

Customers of mobility-on-demand systems want to find vehicles available at pickup points whenever they walk up to them to initiate trips, and they want to find parking spaces available at drop-off points whenever they arrive to drop vehicles off. Assuring this requires sufficient numbers of vehicles and parking spaces to meet demand, appropriate distribution of parking spaces throughout the system, and careful management of vehicle and parking space inventories at access points.

It may not, however, be strictly necessary to have vehicles and parking spaces immediately available upon customer arrival. More generally, operators of mobility-on-demand systems must keep mean waiting times to pick up vehicles, and the variances on those times, within acceptable bounds. Similarly, they must control the means and variances of waiting times to park vehicles. Waiting times extend door-to-door travel times, and waiting times that are highly variable make these times unacceptably unpredictable.

When there are waiting times, queues form—of pedestrians waiting for cars, and of cars waiting for parking spaces. Taxi queues have shown that if they are not too long, pedestrian queues are fairly easy to manage on sidewalks and in parking structures. Vehicle queues, however, often take the form of traffic congestion created by vehicles driving around looking for parking, so they should be avoided.

Using Dynamic Pricing for Balancing

Since demand for vehicles and parking space fluctuates, there will always be a tendency for the system to become unbalanced, with more vehicles than necessary but too few parking spaces at some locations, and too few vehicles but more empty parking spaces than necessary at others. To keep pickup and drop-off queues within acceptable limits, some sort of system-balancing strategy is necessary.

Figure 8.9 illustrates spatial imbalance of supply and demand. At any moment, there will tend to be some areas of the city where vehicle supply exceeds demand, other areas where vehicle supply is insufficient to meet demand, and yet other areas where vehicle supply and demand are nicely balanced. A system operator should try to minimize the areas

Figure 8.9
Spatial imbalance of vehicle supply and demand, over an urban area, in a mobility-on-demand system.

Unsatisfied demand   Balanced supply and demand   Excess supply of excess supply (which waste vehicle capacity that could be put to good use elsewhere), minimize the areas of unsatisfied demand (which leave customers frustrated), and maximize the areas of balanced supply and demand.

Figure 8.10 illustrates temporal imbalance of supply and demand. At any particular access point, both supply and demand will fluctuate over time as customers requiring vehicles arrive and as vehicles arrive and depart. Ideally, the supply and demand curves would match, but in practice they will tend to be out of balance at some moments and in balance at others. A system operator should try to minimize the times when the access point is out of balance and maximize the times when it is balanced.

One approach to system balancing is simply to provide a large margin of safety—many more vehicles and spaces than normally necessary—to absorb demand fluctuations. If this margin is sufficiently large, then vehicle and space stocks will never be driven down to unacceptable levels at any location. But this increases space and vehicle costs, particularly in contexts when fluctuations are large. Instead, the management goal should be to control pickup and drop-off queues without incurring excessive space and vehicle costs.

Another approach is to pick up vehicles at locations where they are not currently needed and move them to locations where they are. This is popular in urban bicycle-sharing systems, such as the Vélib system in Paris, where bicycles are simply loaded onto trucks for transfer. It is also used in supermarket shopping-cart and airport luggage-cart systems, where dispersed carts are collected to form trains for economical transfer back to dispensing racks. And it is sometimes resorted to in trucking and airline fleets, where empty vehicles must occasionally be moved to where they are more useful.

Although this process could be possible for USVs as well, it is a costly process nonetheless. The cost-effectiveness of mobility-on-demand systems depends on reducing this sort of vehicle movement to the minimum. Instead of using a driver to move each automobile, for example, virtual towing of trains of empty vehicles might be used to achieve economies of scale. (Virtual towing is achieved by coordinating the electronic control systems of vehicles rather than physically connecting them as with carriages in a train. Thus one driver can move many vehicles.) With autonomous driving, vehicles might move automatically, in the small hours of the morning, to desired destinations. And mathematical optimization techniques similar to those employed in trucking and airline fleet management can be applied to achieve system balance with minimal vehicle movement.

Taking advantage of trip origin and destination elasticity, combined with price incentives, provides yet another, more elegant strategy. Not all trip origins and destinations are elastic: if you travel from your home to a commuter rail station in the morning, for example, you have little choice about where and when to begin and end your trip. Often, though, you have the flexibility to walk a block or two to find an automobile if there isn't one available right outside your door. Similarly, it might not be a problem

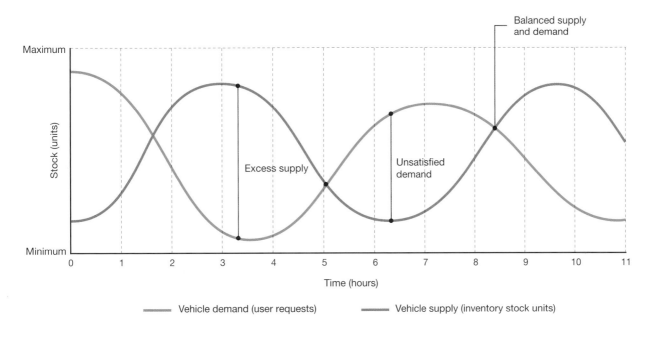

Maximum

Stock (units)

Minimum

0   1   2   3   4   5   6   7   8   9   10   11

Time (hours)

Excess supply

Unsatisfied demand

Balanced supply and demand

——— Vehicle demand (user requests)   ——— Vehicle supply (inventory stock units)

Figure 8.10
Temporal imbalance of vehicle supply and demand at
an access point in a mobility-on-demand system.

for you to park at a little distance from your destination. And, if slightly less convenient origins and destinations result in lower-priced rentals, you will have an incentive to use them even if more convenient pickup and drop-off points are available.

So, just as price incentives can smooth peak demands for electricity, road space (through congestion pricing), and parking space, they can also more smoothly distribute demands for shared-use automobile pickup and drop-off, enabling close matching of supply to demand while minimizing costly overcapacity and movement of empty vehicles (figure 8.11). In other words, dynamic pricing of vehicle possession can effectively keep the spatial and temporal distribution of shared-use vehicle and parking space supply in equilibrium with the demand.

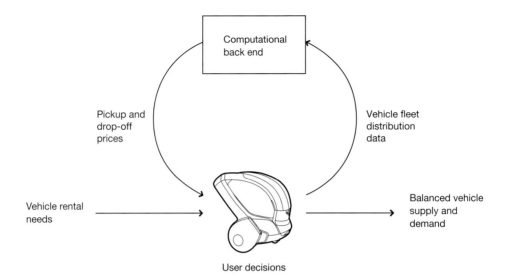

Figure 8.11
The logic of dynamic mobility-on-demand pricing: price signals regulate the demand for vehicles and parking spaces—smoothing the spatial distribution of demand across the service area and the temporal distribution of demand at each access point.

In a perfectly balanced mobility-on-demand system, there would be just one parking space at each pickup and drop-off location. Whenever a customer arrived at a location to pick up a vehicle, a vehicle would show up at exactly the same time. And, whenever a customer arrived at a drop-off location to park, there would be just one parking space available. Obviously this is impossible in practice, so pickup and drop-off locations provide stocks of vehicles and parking spaces awaiting use. From the system operator's perspective, the goal is to minimize those stocks (since vehicles and parking space cost the operator money) while assuring sufficient levels of vehicle and parking-space availability for good customer service.

From the customer's perspective, the goal is to minimize trip cost (figure 8.12). This is accomplished by locating a nearby, relatively inexpensive pickup point—perhaps using a smart phone or other mobile device, walking to it and picking up a vehicle, choosing a relatively inexpensive drop-off point near the destination, driving to the drop-off point, and finally dropping off the vehicle and walking the rest of the way.

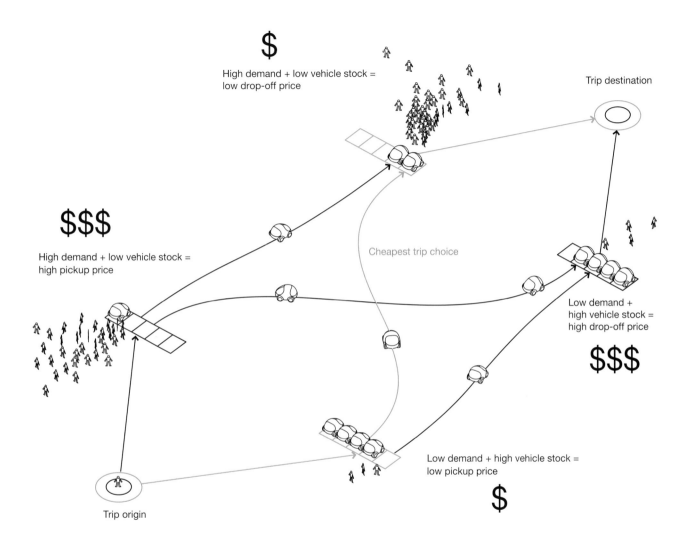

$

High demand + low vehicle stock =
low drop-off price

Trip destination

$$$

High demand + low vehicle stock =
high pickup price

Cheapest trip choice

Low demand +
high vehicle stock =
high drop-off price

$$$

Low demand + high vehicle stock =
low pickup price

$

Trip origin

Figure 8.12
Customers of mobility-on-demand systems seek
to minimize total trip costs.

Combined Pricing

With these various sorts of markets in operation, the total cost of a trip is the cost of the electricity consumed plus the costs of road space, parking, vehicle use, and insurance.

But these costs are not independent: there are some important interactions to consider. It is, for example, advantageous to be parked and buying electricity when electricity prices are low, and disadvantageous to run low on battery charge and be forced to recharge when prices are high. It saves money to be off the road when congestion is high, and—since demand for parking is lowest when most cars are on the road—that is also when it's likely be cheapest to find parking, provided that you are willing to settle for a space in a sparsely occupied part of the system. Seeking parking in congested areas, though, is likely to incur not only high parking costs, but also high road costs on congested approaches.

There is, as well, an opportunity to integrate insurance into overall trip cost calculations. Rates can be tailored to the time of day and location, based on actuarial data. In the end, sufficiently sophisticated software—taking account of all the cost components and their interactions—can calculate the total cost of an automobile trip from a specified origin to a specified destination, at a specified time, and express it as a single number. Drivers can choose among variously priced destination and schedule options, much as air travelers now shop for fares (figure 8.13).

Today, by contrast, motorists choose trip times and routes based only on the most rudimentary understanding of the associated costs. They do not get useful price signals, and they do not have effective incentives for resource-conserving behavior or strategies for acting on such incentives. We can remedy these problems by creating dynamic, electronically managed markets for electricity, road space, parking space, and vehicle access, and by providing the wireless communication capabilities and onboard intelligence necessary for drivers to respond appropriately to price signals.

The effects of trip pricing will vary with context. Where energy is costly, this may dominate overall trip prices. In highly congested urban areas, road and parking space prices may dominate. Sometimes the vehicle component of prices may be particularly significant—although low-cost electric-drive vehicles and shared use can drive this down to much lower levels than those prevailing with today's cars. People generally value their time highly, and they also value mobility predictability—the ability to arrive at destinations on time and keep to their schedule—so it is a reasonable hypothesis that many will be willing to pay premiums for trips that offer short and certain travel times, and generally will need significant discounts to accept travel times that are longer and less reliable. This, however, will vary with income; price discounts that are trivial to the wealthy may be decisive to the less well off. And competition from other mobility modes—walking, taxis, public transport, and so on—will set limits on price variations.

Autonomous operation could become a game changer as it allows the driver to make more productive use of travel time and may make it more

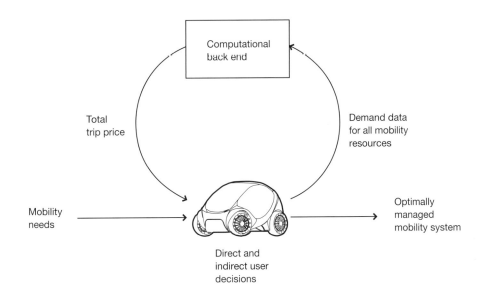

Figure 8.13
Total trip costs.

acceptable to take a longer, less costly route. It also frees the driver to receive information from the outside that normally would be a distraction and a cause for safety concerns.

Integrating Search and Advertising

As we have already seen, the ability of drivers to respond to price signals depends on what they need to do. Some car trips, like visits to your mother's home, have unique destinations. Some, like journeys to the theater for curtain time, have inflexible times. Some have both. But cities, generally, offer choices: if you want to go to a supermarket, for example, there are probably several within reasonable travel time, and there will be more if you can travel faster or are prepared to travel longer. You can choose among supermarkets not only according to their suitability for your shopping purposes, but also according to the travel times and costs of accessing them.

The existence of choices among trip destinations and times provides some of the elasticity in mobility demand that is needed for keeping supply and demand in balance through pricing. But it also raises a question: How can urbanites discover and evaluate the options that are available to them?

The traditional answer is that people learn to access the resources of their cities through experience, augmented on occasion by maps and directories such as the Yellow Pages. But GPS navigation systems can now provide an additional answer through the integration of urban search engines.

These capabilities are analogous to those of search engines that provide access to the resources of the World Wide Web. They already exist in rudimentary form in many GPS navigation systems, and there is great potential to enhance their capabilities. If you want to go to a supermarket, your vehicle's navigation system should show you the supermarkets that are accessible within a specified time or at a specified cost, and will be open when you get there. Furthermore, you can sometimes avoid a trip if opening times, availability of stock, best local prices, and so on are known ahead of time.

This also provides an opportunity for location-based advertising. As with Google, advertisements matched to search topics might appear with the search results. Advertisers might buy visibility within this sort of system. And they might also subsidize travel costs to their sites.

The combination of sophisticated road and parking space markets with urban search and location-based advertising opens up the possibility of some interesting new business models for personal urban mobility. Currently, the responsibility for identifying destinations and paying for travel to them rests primarily with drivers. In the future, advertisers might take over much more of that responsibility.

Summary: Automobiles as Interfaces to the City

Within electronically managed, dynamically priced personal urban mobility systems, smart, connected, location-aware vehicles serve as interfaces to the city, much as the combination of a browser and a search engine serves as an interface to the World Wide Web. They help their drivers locate the resources that they want to access within the urban fabric, advise them on how to most conveniently and efficiently reach these resources—taking account of trip time, energy cost, road-space cost, parking space cost, and vehicle cost—and guide them along chosen routes from trip origins to destinations.

This implies a gradual transfer of information display from the external urban environment to automobile dashboards. Advertising is much better located on dashboards or the screens of smart phones than billboards—where it doesn't clutter the cityscape, where it can become dynamic and location-sensitive, where it can be integrated with search and navigation systems, and where it can be personalized. Speed limit information is better displayed on speedometers than on static road signs so that it can be tailored to the vision capabilities and preferences of the driver and where it can be compared directly to current vehicle speed. Eventually, even stop signs and traffic lights might become more dynamic (a vehicle shouldn't have to stop at an intersection if no vehicles are coming from other directions) and move onto the dashboard. And if vehicles can drive autonomously, then completely new types of information can be displayed on dashboards to entertain and inform the "driver."

The result of all this is a different role for the dashboard and a radically different driver experience—both when the driver wants to grab the wheel, and when the vehicle operates in autonomous mode. Early automobiles were fractious and unreliable contraptions, so dashboards evolved as interfaces to their engines. They told drivers about gasoline, oil, water, battery charge, speed, revolutions per minute, and so on. In simple, reliable electric-drive cars, though, most of this is no longer necessary. Dashboards can become what they really need to be: map-based, rather than dial- and gauge-based interfaces, that help drivers to safely, efficiently, and conveniently make use of what their cities have to offer.

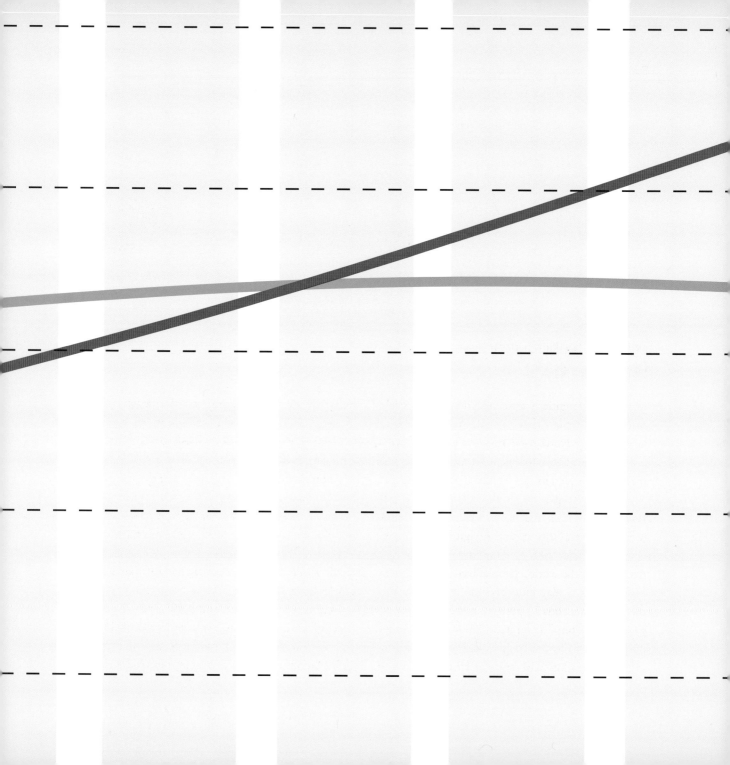

Personal Mobility in an
Urbanizing World

In the early twentieth century, when the horseless carriage was replacing the horse, the America that greeted Henry Ford's Model T was a nation of farmers. And it had access to vast domestic oil reserves. It made sense to design cars for those conditions.

Now, though, the rural–urban ratio has reversed (figure 9.1), and the requirements for automobile and mobility system design have changed with it. The year 2007 was the first year in history that more of the world's people lived in urban than in rural areas. Today, cities account for 2 percent of the world's area but 75 percent of the world's energy consumption. The United Nations projects that 60 percent of the world's population will be living in urban areas by 2030 and that 80 percent of the

world's wealth will be concentrated in urban areas by that date. The increasing concentration of population and wealth to cities is likely to continue—especially in the developing world.

Forecasts of future energy demand are driven by automobile sales growth, which will take place primarily in cities in the developing world. Given that the population and wealth increasingly reside in urban areas, it is clear that vehicle miles traveled (VMT) will increasingly accrue in urban areas, as evidenced in the United States over the last half a century (figure 9.2). Congestion will exacerbate the effect of urban vehicle miles traveled on overall energy consumption. It will not be possible to tackle global automotive energy consumption and green-

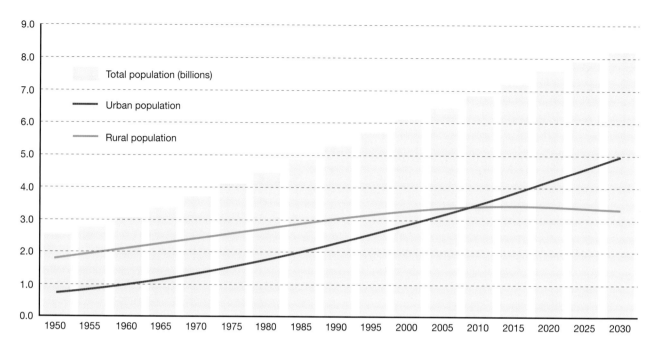

Figure 9.1
Urban, rural, and world population, 1950–2030.

Chapter 9

house gas emissions effectively without rethinking urban mobility.

In this chapter, we show that electric-drive USVs can provide better, cleaner, more efficient personal mobility within cities than today's gasoline-powered cars. Switching to these vehicles will produce dramatic improvements in urban sustainability and energy security. Doing so will create an attractive new market, grounded in green technology, for the automobile industry. And further efficiencies and opportunities will result from utilizing these vehicles within systems for dynamic pricing and electronic management of electric grids, road space, parking space, and vehicle fleets, as discussed in previous chapters.

In the cities of the developed world, where there are already intense concentrations of automobiles, the opportunity exists to gradually replace gasoline-powered vehicles with electric-drive vehicles. The situation is different in the rapidly growing cities of the developing world. Here, it will be particularly important to respond in new ways to the skyrocketing demand for personal mobility and the ability to pay for it. China, for example, is already leading the way in the production and use of electric bicycles and scooters, and is giving indications of taking the same path with electric-drive automobiles.[1]

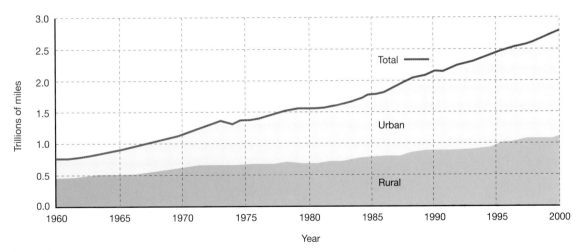

Figure 9.2
Urban driving is increasing: vehicle miles traveled 1960–2000.

Vehicle Ownership, Personal Wealth, and Population Density

Observed data show that levels of vehicle ownership increase with personal income at a national level (figure 9.3) and with GDP per capita (figure 9.4). It generally decreases with urban population density (figure 9.5). In particularly densely populated cities, such as New York and Singapore, however, vehicle ownership is sometimes quite low because of congestion, parking costs, high vehicle taxes, and the inconvenience of unpredictable travel times.

Moreover, as population density increases it becomes easier to justify investment in public transportation systems. This, in turn, can reduce automobile usage, as shown in the case of Paris (figure 9.6),

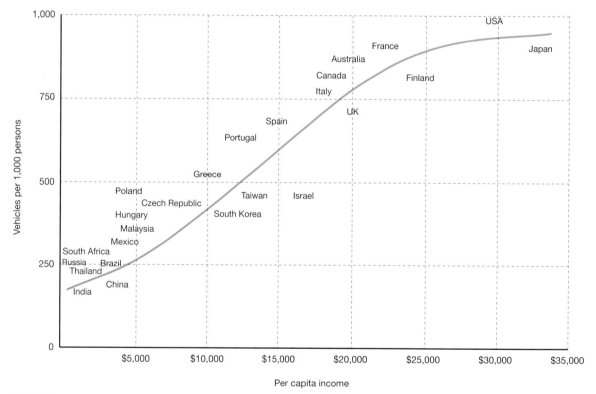

Figure 9.3
Vehicle ownership typically increases with per capita income.

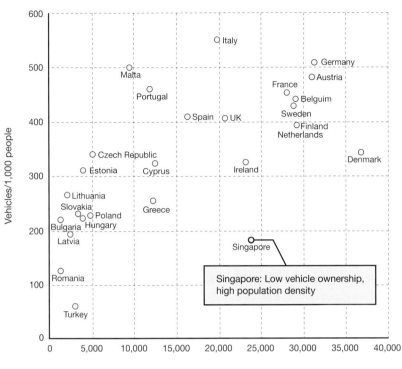

Figure 9.4
Vehicle ownership increases with
GDP per capita.

Metropolitan area	Population density	Vehicles/ household
Atlanta		1.66
Dallas		1.59
Detroit		1.57
Washington DC		1.56
Los Angeles		1.54
Houston		1.53
Chicago		1.45
Boston		1.42
Philadelphia		1.36
New York		0.7

Low density

Medium density

High density

Figure 9.5
Vehicle ownership decreases as population
density increases.

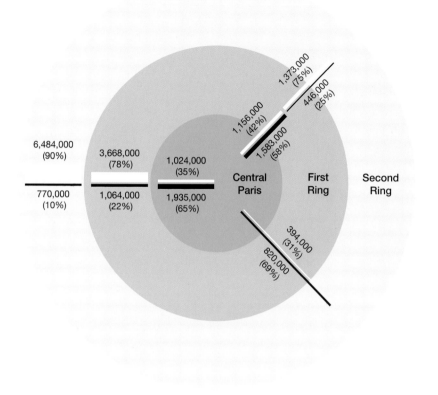

Central Paris =
Arrondissements 1–20

First Ring =
departments of Hauts-de-Seine,
Seine-Saint-Denis, and Val-de-Marne

Second Ring =
remainder of the Île-de-France region

Trips involving central Paris
Share of total trips = 33.4%
Public transport's share = 62.8%

Trips not involving central Paris
Share of total trips = 66.6%
Public transport's share = 16.5%

Total trips by car 14,099,000 (68%)

Total trips by Public Transportation
6,618,000 (32%)

Figure 9.6
Daily trips by car and by public transportation
in and around Paris. There is high population
density and low automobile use at the center,
but the reverse at the periphery.

which has high population density and low automobile usage at the center, but the reverse at the periphery. Singapore, a city-state with exceptionally high population density, has an explicit policy of making public transportation the mode of choice. By using a combination of carrots (secure, comfortable, affordable, convenient public transportation) and sticks (registration taxes on vehicle purchase, congestion charging), it maintains a much lower rate of vehicle ownership than its per capita wealth would predict.

In the emerging markets of the developing world, the population density of major cities is considerably higher than that in Europe and much higher than for cities in North America (figure 9.7). If past trends are any indicator, the likelihood of lower conventional automobile sales in major growth markets (where wealth resides primarily in the major cities) is high. In any case, high conventional automobile rates in these cities would create immense energy and environmental problems.

Figure 9.7
Population densities for the world's twenty largest cities.

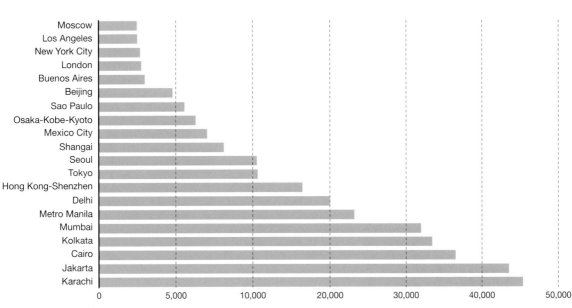

People per square mile

The crucial question, then, is how best to provide city dwellers with personal mobility in the future. The solution must combine effective responses to urban challenges (energy, environment, safety, congestion, parking) with vivid appeal to consumer aspirations for vehicle ownership or access (affordability, fun driving experience, style, personalization, comfort, privacy, utility, and so on).

To answer this question effectively, it is vital to recognize several conditions that characterize personal urban mobility: urban trips are short; urban driving speeds are low; urban traffic congestion diminishes throughput and energy efficiency; parking competes with other uses of urban space; automobile use in cities has a variety of negative externalities; and there is increasing pressure to limit or eliminate automobile use within cities. In the following sections we discuss and quantify these conditions.

Urban Trips Are Short

Throughout history, people have tended to allocate approximately 60–90 minutes per day for mobility (figure 9.8). Centuries ago, this placed the typical geographic span of a city at around 3–4 miles (or the distance that could be walked reasonably in

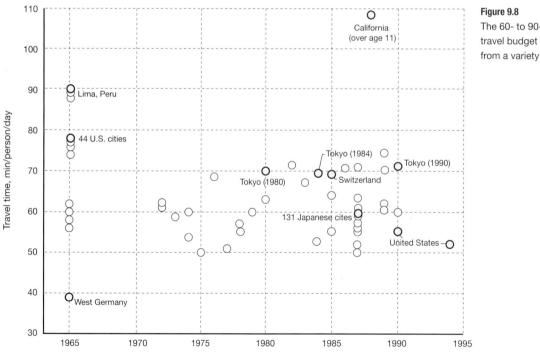

Figure 9.8
The 60- to 90-minute daily travel budget illustrated by data from a variety of contexts.

one hour). As faster forms of mobility evolved, cities grew to maintain this informal "rule of thumb." Many big cities today are 20–30 miles in diameter, which is about the distance that can be traversed in one hour with a car. However, because of congestion, it is difficult to travel at these speeds unless one is driving in the suburbs.

Urban car trips are short by comparison with intercity and rural trips. The growing concentration of population in cities, then, translates into generally shorter automobile trips and lower driving speeds than those occurring within the more dispersed settlement patterns of the past. Figure 9.9, for example, illustrates daily driving ranges in the United States. It indicates that more than half of Americans travel less than 20 miles a day—suggesting that a 25-mile-range battery-electric vehicle, recharged at work to give a 50-mile daily range, might be sufficient for 75 percent of Americans (and people living in other countries tend to drive even fewer miles). Even when areas with longer commutes are included, 94 percent of Americans commute less than 80 miles per day, and 98 percent commute less than 110 miles. So the 300-mile ranges provided by today's automobiles are rarely needed.

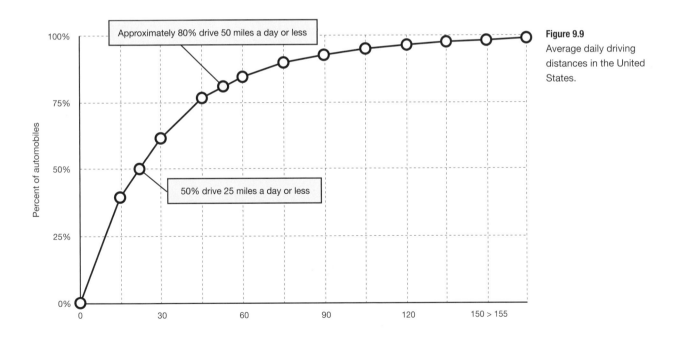

Figure 9.9
Average daily driving distances in the United States.

Urban Driving Speeds Are Low

Figure 9.10 illustrates the widely observed relationship between population density and average road speed. In high-density urban areas, traffic speeds can be less than 10 miles per hour, and in most cities they remain within the 15- to 25-miles-per-hour range. Under these conditions, the 100-mile-per-hour top speeds of today's cars are of very little use.

Furthermore, the number of passengers transported along a road in an hour—its throughput—depends on *average* vehicle speed and spacing, not top speed. As we will show later, even these relatively low average speeds can allow higher throughputs than those achieved by today's urban streets and roads. When throughput under conditions of high mobility demand is the goal, it is more important to design vehicles and infrastructures for moderate but consistent speed than for high top speed.

Vehicles with limited top speeds, as appropriate for urban conditions, need not be sluggish and dull to drive. Lightweight electric-drive vehicles have high-torque motors and less mass to accelerate. They can be designed to provide spirited launches and lively handling at low to moderate speeds without incurring the costs and inefficiencies of 250-horse-power gasoline engines. Moreover, the highly prized maneuverability enabled by robot wheels will also make an electric-drive USV fun to drive.

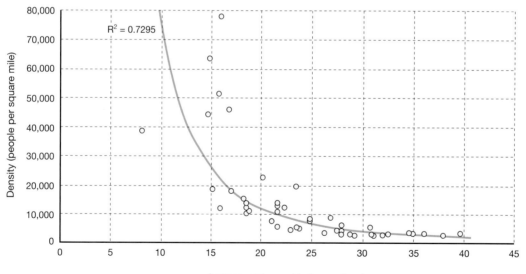

Figure 9.10
Average driving speed drops as population density rises.

Traffic Congestion Diminishes Throughput and Energy Efficiency

Every city in the United States has seen an increase in congestion over the last few decades, as vehicle miles traveled have increased at a higher rate than new road construction. Recurring congestion (commonly associated with rush-hour traffic) makes up about 40 percent of the congestion measured on roads in the United States (figure 9.11). The remaining 60 percent is nonrecurring and is typically related to traffic accidents, road construction, and poor weather. Figure 9.12 shows that there has been a slow but steady increase in the fraction of time that drivers spend in congested traffic in recent years.

Only with the 2008–2009 economic downturn has congestion decreased in recent memory. It decreased by 30 percent in the United States, with only a 3 percent reduction in VMT compared with the previous year. This illustrates two important points about congestion: it goes hand in hand with economic activity, and it is highly nonlinear, with a tipping point that can be triggered by the addition or subtraction of a relatively small number of vehicles. Efforts to make driving easier, for example by building new roads, will only temporarily reduce congestion although it will increase throughput. Typically, within a few months, more demand will have been induced, and traffic will have returned to the same congested levels as before.

The congestion in other parts of the world is often much worse than it is in the United States, partly because of the limited land space. Cities like

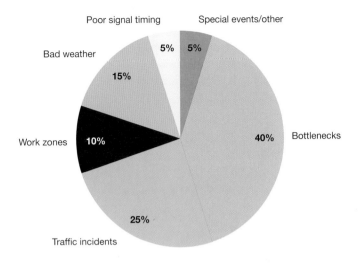

Figure 9.11
Sources of traffic congestion in the United States.

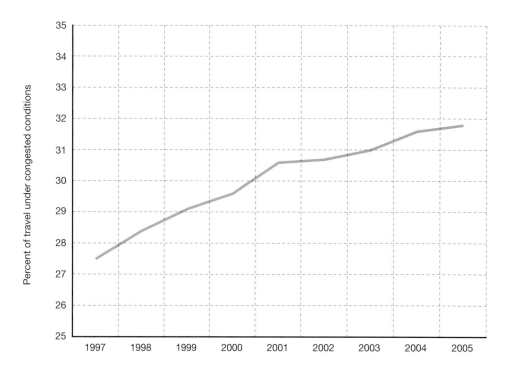

Figure 9.12
Travel under congested conditions is increasing
in the United States.

Bangkok and Mumbai and even parts of Manhattan are notorious for having traffic speeds that are barely above walking pace.

This creates a pressing need for the congestion reduction measures that we have introduced: reducing the footprints of vehicles, smoothing traffic flow, dynamically pricing road space according to congestion levels, and providing mobility-on-demand systems as an alternative to private automobile ownership. These measures may be applied individually or in various combinations.

Parking Competes with Other Uses of Urban Space

Figure 9.13 illustrates, for the example of Albuquerque, New Mexico, the immense amount of downtown land that may be consumed by parking space. Conversely, as is the case in Manhattan, allocation of little space to parking makes it expensive and difficult to find. Parking space competes with urban land for other uses. This creates a need for smaller vehicles that occupy less parking space, for more efficient configurations of parking space, and for more efficient management of available parking space through occupancy sensing and dynamic pricing.

As automobile ownership and usage in a city grows, so does demand for parking space. If large amounts of parking space are provided in response, this hollows out the city and makes it less attractive for human use. Conversely, if the demand for parking space can be reduced, space that was previously occupied by parking can be turned to other uses.

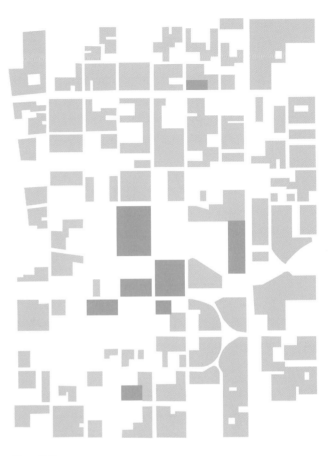

Figure 9.13

Parking space versus building and people space in downtown Albuquerque, New Mexico. The white areas represent streets and buildings, the light areas surface parking lots, and the dark areas multistory parking structures.

Automobile Use Has Negative Externalities

The negative externalities of automobile use are the costs that are borne not by individual drivers, but by society as a whole. The human and monetary costs of traffic accidents constitute one of the most significant of these externalities. We tend to equate road safety with protecting passengers in vehicles; but there is also a tremendous need to reduce accidents involving more vulnerable road users, such as cyclists and pedestrians, since these comprise the majority of road-related fatalities in the urban environments of the developing world. Lighter-weight, lower-speed vehicles with sophisticated crash avoidance capabilities are part of the solution to this problem.

The characteristics of gasoline-powered automobiles result in further negative externalities. Slow traffic speeds in cities are caused by a variety of factors, such as limited road and parking space and high accident rates. They contribute to higher energy consumption because a traditional automobile's fuel economy deteriorates at low speeds. In the limit of reduced speed, when the vehicle is at a standstill, it is not traveling forward but is still consuming energy (air conditioning use, for example) and is, therefore, "achieving" zero miles per gallon.

Figure 9.14 illustrates this point. The average speed over the Federal Test Procedure (used for regulating the fuel economy of automobiles in the United States) is 19.6 mph. Any vehicle's fuel economy, measured on this cycle, will be much lower than what it will be if it is traveling at the same average speed but on "cruise control." The loss is due to

the transient (stop-start and idle) nature of the drive cycle, which is intended to mimic real-life driving and which can only partly be addressed with regenerative braking on gasoline-electric hybrid vehicles. (Regenerative braking is reversing the direction of the electric motor to act as a generator when the vehicle decelerates, in order to feed current back into the battery and increase its state of charge, which increases fuel economy.)

As we saw in chapters 3 and 8, smoothing traffic flow is one of the capabilities provided by the Mobility Internet: if vehicles can communicate with each other, they can coordinate with each other and arrange themselves to avoid accidents and gridlock. The Mobility Internet has great potential to improve the energy efficiency of each vehicle affected by surrounding traffic (potentially a multiplier effect of hundreds compared with improving the efficiency of a single vehicle).

Raising the average speed of traffic from 10 mph to 15 mph, for example, could increase the fuel economy of each vehicle in the traffic by around 20 percent. Smoother traffic and higher average traffic speeds will also lessen the transient tailpipe emissions that contribute to smog, which can be a major health issue in some cities.

Battery-electric vehicles do not, of course, produce tailpipe emissions. But, like traditional automobiles, they are more energy efficient when they conserve momentum by traveling within smooth traffic streams at optimum speeds.

Urban traffic noise is yet another negative externality of traditional automobiles in cities. This

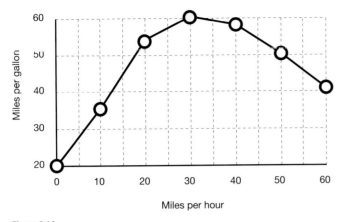

Figure 9.14
There is an optimum constant speed for energy efficiency.

makes heavily traveled streets unpleasant, and often creates zones of blight and reduced property value around motorways and freeways. Property owners often respond to traffic noise problems by sealing windows, which necessitates air conditioning even where natural ventilation would be more pleasant and energy efficient. But electric motors, by contrast, are almost silent, so the introduction of electric USVs at a large scale will create the opportunity to reverse this trend.

Cities Are Limiting Automobile Use

Given these negative externalities and other challenges, it is not surprising that cities around the world are experimenting with making alternatives to

the automobile more appealing and deterring people from driving a car. One approach is to tax the externalities, so increasing the cost of driving. Another is to introduce some combination of direct limits on automobile use and provision of alternatives.

Measures to reduce automobile use can take the form of installing dedicated bicycle and bus lanes to make traveling by bicycle safer and make traveling by bus faster. The Brazilian city of Curitiba, for example, has pioneered the use of Bus Rapid Transit (BRT) systems that provide buses with dedicated lanes and traffic light priority. Another option is to promote novel sharing schemes for bicycles and cars, as in Paris. Or driving in certain areas can be made more expensive through road pricing and electronic tolls. In general, cities recognize that to compete internationally they need to offer a high quality of life, and this can lead to them becoming even more proactive than national governments in regulating vehicles.

Motivated by increasingly urgent concerns about greenhouse gas emissions, air pollution, and congestion, many major cities may soon permit only zero-emissions vehicles such as electric USVs or may even consider banning automobiles completely. The assumption that many city governments are making is that the automobile as we know it today is not compatible with their vision for sustainability, which embraces a high quality of life through making their roads safer, cleaner, and quieter, and by making transport available to all at an affordable cost.

As an example, let us consider New York City, where car ownership is in decline. There were 200,000 fewer registered drivers in 2006 compared with five years earlier, a decrease of over 10 percent. NYC has a vehicle ownership rate that is less than half that for the United States as a whole. Not surprisingly, people are less likely to own a vehicle if they live in densely populated areas; only 1 in 7 residents in Manhattan owns a vehicle, compared to nearly half of Staten Island residents (the least densley populated of New York's boroughs). In Manhattan, only 2 percent of households have more than one vehicle, 20 percent have one vehicle, and the remaining 78 percent of households have no vehicle.

According to the 2001 National Household Travel Survey, two-thirds of all trips are 30 minutes or less and the average occupancy of a vehicle is

1.81. Average traffic speeds across Manhattan are below 10 mph, but as can be seen in figure 9.15, the average speed of cars in NYC (averaged across all the boroughs and over an entire day) was 18.8 mph in 2001, down from 24.3 mph in 1995. During the same time period, the average number of miles driven by car each day fell from 12.2 miles to 9.9 miles. Nevertheless, low as these numbers are, they are still faster and longer than for transit systems.

Motor vehicles account for 83 percent of all the transportation energy consumed in NYC (electric trains constitute nearly all of the rest) and 100 percent of the oil, which amounted to 741 million gallons in 2005.

In the next section, we illustrate and quantify some of the ways in which cities can provide superior levels of personal mobility while minimizing the negative effects of urban automobile use, through utilization of USVs and electronic coordination of vehicle movement.

Safety Improvements with USVs

A central goal when introducing any new mobility system to a city should be to improve the overall safety of the city. This translates into two objectives: providing an appropriate level of safety to vehicle occupants and improving the safety for vulnerable road users (pedestrians, cyclists, motorcyclists, and so on).

With a USV (plus two occupants) weighing around 1,000 lbs and typically traveling at speeds below 25 mph there is a significant reduction in

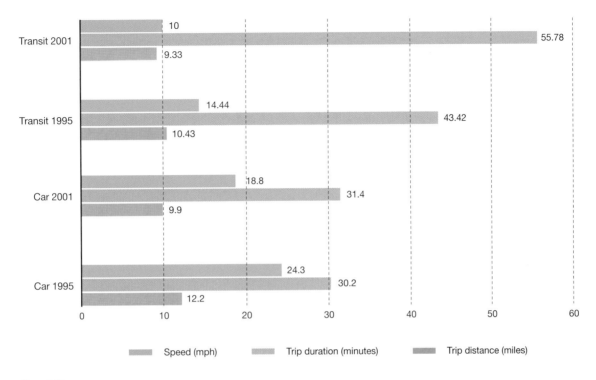

Figure 9.15
Mobility trends for New York City.

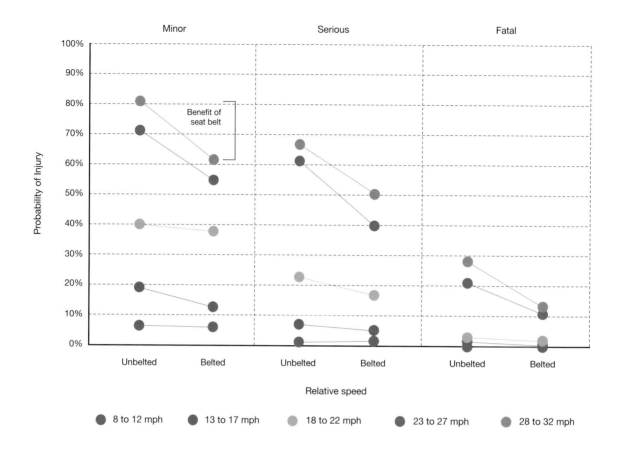

Figure 9.16
Data from vehicle-to-vehicle crashes show, with
seat belt use, high probability of surviving collisions
below 15 mph.

kinetic energy to begin with, compared with a conventional vehicle that might weigh five times as much and could be traveling (briefly) at much higher speeds, even in built-up areas.

The key to improving the safety of USV occupants, in addition to perhaps operating on dedicated lanes, is to use sensing and wireless communications to avoid or mitigate collisions with other vehicles. If a collision is imminent at traffic speeds of around 25 mph then there will be time to decelerate, so that impact speeds will nearly always be below 15 mph. Data from vehicle-to-vehicle crashes that occur in the existing vehicle fleet suggest that at speeds below 15 mph, and with seat belt usage, nearly all collisions should be survivable (figure 9.16).

Injury probabilities for vehicle-to-pedestrian crashes from the existing fleet suggest that pedestrian injuries may also be significantly less severe if the vehicle can decelerate to below 15 mph (figure 9.17). If initial vehicle speeds are close to 25 mph then this can be achieved today with a combination of sensing and fast-acting brake systems. However, the system will be more robust and less expensive in the near future, as vehicles could also be able to communicate wirelessly with people. As noted in chapter 2, for example, General Motors has already prototyped a wireless transponder the size of a handheld device that can warn drivers and pedestrians (or cyclists) of each other's presence even if either is hidden or distracted.

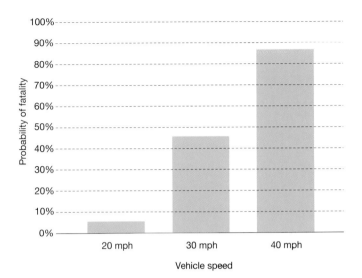

Figure 9.17
Fatality probabilities for vehicle-to-pedestrian crashes.

Energy Efficiency Goals Achievable through USVs

Electric-drive USVs can be compared with bus, Bus Rapid Transit (BRT), conventional automobiles, and Neighborhood Electric Vehicles (NEVs). It is no surprise that, by virtue of their lighter vehicle mass, USVs are more energy efficient than the other personal mobility options; but what may be surprising is how well they compare with public transportation, especially when we take into account realistic or average seat occupancies (figure 9.18). A standard bus will have to be fully seated with 44 occupants to use approximately the same amount of energy per passenger as 22 electric-drive USVs with two occupants in each of them. Even in this case, the greenhouse gas emissions, tailpipe emissions, and oil imports are not equal, since the bus typically runs on diesel whereas electric-drive vehicles run on electricity and can derive their energy from a wide variety of sources. Moreover, as we've seen, the percentage of electricity generated from renewable sources could even be stimulated by the commercialization of electric-drive vehicles.

From a cost perspective, one bus can cost as much as $400,000. USVs, on the other hand, could cost under $10,000 and may typically accommodate 1.8 people. For the same price as a bus it should be possible to purchase over 40 USVs, which can accommodate as many as 80 people. Although BRT offers higher throughput and higher occupancy loads than standard buses, BRT buses are also significantly more expensive and have slightly lower energy efficiency per passenger mile than standard buses.

A fleet of 20 to 40 USVs also provides far more flexibility than a bus, which has a fixed route and schedule. As we saw in chapter 8, though, USVs in mobility-on-demand systems can provide service when and where it is needed.

Throughput Improvements

BRT is being applied in many cities around the world as a lower cost alternative to subway systems and elevated railways while still offering similar throughputs. NYC has the Lincoln Tunnel BRT system that can move 25,000 people per hour during rush hour. One bus leaves every 5 seconds, but there are enough buses in the terminal at any one time to allow 3–4 minutes for passenger loading and unloading.

If two USVs are placed side by side with a 50cm lateral separation, the width of the lane will need to be increased to 3.6m (as compared with 3m for a standard bus lane). With dedicated short-range communications, it is possible to allow a platooning separation of around 6.7m today; and further advances in V2V should allow this to be reduced to 4.4m in the near term and 2.2m in the future (figure 9.19). If USVs run at 25 mph with a platooning gap of 2.2m, or 7 feet, it is possible to surpass the throughput of the Lincoln Tunnel Bus, assuming an occupancy of two passengers per vehicle and two vehicles side by side

Energy efficiency and emissions	Dimensions (L x W x H)	Curb mass (kg)	Price ($)	Passenger capacity	Average seat occupancy	Gasoline equivalent fuel economy (mpg)	Well-to-wheel energy consumption (megajoules/ passenger-mile)	CO_2 emissions (grams/ passenger-mile)
Standard transit bus	40' × 102" × 135"	14000	400,000	44 seated + 34 standing (78 total)	24	4	1.56	100
Articulated BRT bus	60' × 100" × 134"	21500	980,000	27 seated + 90 standing (117 total)	35	2.4	1.79	115
Prius	175" × 68" × 58.7"	1330	22,000	5 seat, WB=106.3"	2	48	1.61	118
Smart 4 two	106.1" × 61.4" × 60.7"	820	13,990	2 seat, WB=73.5"	2	33	2.33	168
Chevrolet Malibu	191.8" × 70.3" × 57.1"	1550	19,900	5 seat, WB=112.3"	2	22	3.45	281
Chevrolet Malibu hybrid	191.8" × 70.3" × 57.1"	1630	23,900	5 seat, WB=112.3"	2	26	2.94	235
GEM-e2	99" × 55" × 70"	517	7,395	2 seat, WB=72"	2	230	1.01	65
Zenn	121" × 58.8" × 55.9"	636	12,000	2 seat, WB=81.9"	2	224	1.04	63.5
300 kg USV	59.4" × 51.5" × 67.3"	300		2 seat	2	369	0.63	39.6
450 kg USV	59.4" × 51.5" × 67.3"	450		2 seat	2	263	0.88	54.5

Standard transit bus   Articulated BRT bus   Chevrolet Malibu   Zenn

Figure 9.18
Comparison of various modes of urban transport.

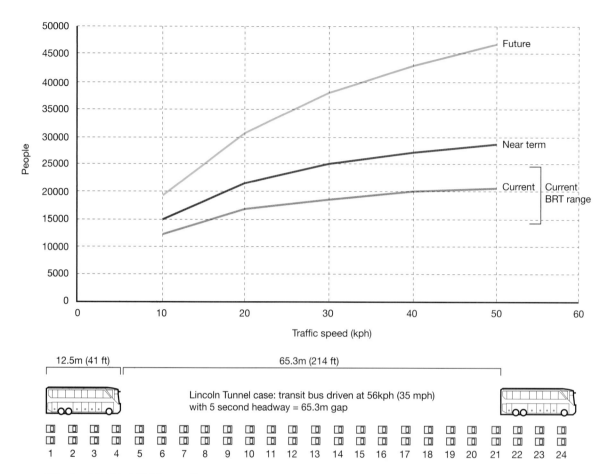

Figure 9.19
Throughput comparison between BRT and USVs.

Parking Space Reductions

The dramatic reduction in vehicle size may not matter much in low-density suburban and rural areas, but it is of great benefit in densely populated urban areas where potential parking space is scarce and land values are high. As the UCLA economist Don Shoup has shown in his book *The High Cost of Free Parking*, the cost to cities of providing street parking is much higher than generally realized.[2] The cost to real estate developers of providing parking lots and structures is also very high. Reducing parking space requirements in urban areas can reduce street congestion, reduce the overhead imposed on urban life by parking costs, and free up valuable land for more productive uses. In heavily built-up urban areas that are underserved by parking and do not provide opportunities to create additional parking space, the only way to park more cars is to increase parking density.

In a typical Manhattan block (ignoring loading zones, fire hydrants, and so on) it is possible to accommodate around 80 cars. The USV's footprint is significantly smaller than a typical car, so it should be possible to increase this number to around 250 (figure 9.20). What's more, USV occupants can exit the vehicle directly onto the sidewalk, as the vehicle can park nose in to the curb.

In a typical parking lot, only one-third of the land area is actually covered by cars when the lot is "full"; the rest of the space is required for cars to turn (aisle space) or to allow people to get in and out of their vehicle. With higher maneuverability (the ability to "turn on a dime") and automated

parking, both of these space requirements can be reduced. (It could even be possible for the vehicle to spin around inside its space so that the driver can exit facing forward, although with automated retrieval this may not even be necessary.) The combination of smaller footprint, higher maneuverability, and automated parking should allow the size of a parking lot that can hold 100 USVs to be three to four times smaller than a conventional parking lot (figure 9.21). Cities will prize the extra land space freed up by USVs, and drivers could benefit from lower parking prices.

Overall Effects on Urban Space and Civic Amenity

As we have demonstrated, several factors combine to enable electric-drive USVs to reduce the gross parking space requirements of buildings and urban areas very significantly. In part, this reduction is due to the smaller footprint of these vehicles and their minimal need for vehicle-to-vehicle clearance and access aisles. Further space savings result from electronic location of available parking spaces, which enables better matching of supply to demand and less parking space wastage. Finally, where mobility-on-demand systems are employed, these enable much higher vehicle-utilization rates; vehicles are on the road more, so they are in parking spaces less.

The actual reductions that can be achieved depend on the details of use patterns and on the parking layout possibilities afforded by particular sites and urban contexts, but some rough calculations can indicate the magnitudes of possible savings. As

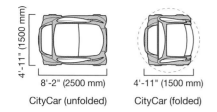

4'-11" (1500 mm)

8'-2" (2500 mm)   4'-11" (1500 mm)

CityCar (unfolded)   CityCar (folded)

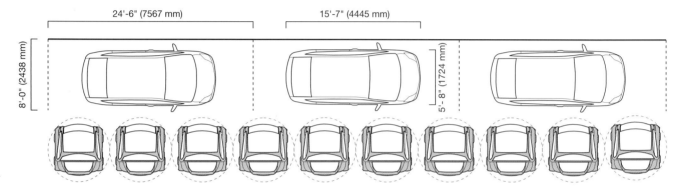

24'-6" (7567 mm)   15'-7" (4445 mm)

8'-0" (2438 mm)   5'- 8" (1724 mm)

Folded CityCar vs. conventional 4-door sedan
Parking ratio = 3.3 : 1

Figure 9.20
Street parking footprint comparison between
USVs and traditional cars.

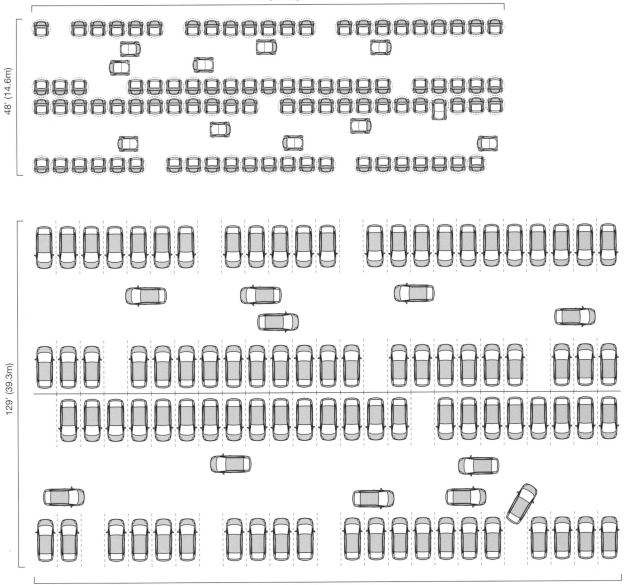

180' (54.9m)

48' (14.6m)

129' (39.3m)

225' (68.6m)

100 CityCars (folded) vs. 100 conventional cars
Parking ratio = 3.4 : 1

Figure 9.21
Parking lot space comparison between
USVs and traditional cars.

we have seen, at least three times as many electric-drive USVs (sometimes more) can be parked in a given area as conventional automobiles. If a mobility-on-demand system reduces the time that an automobile is parked from 80 percent to 20 percent, this reduces the need for parking spaces by another factor of four. (In practice this factor will be somewhat lower, since the demand for vehicles will be distributed unevenly throughout the day, and parking space must be sized to accommodate peak demand. However, mobility-on-demand fleet vehicles can often be "warehoused" at inexpensive, peripheral locations when demand for vehicles is low.) The combined effect is to reduce parking space requirements by at least an order of magnitude.

This reduction is not significant in low-density suburban areas, where the supply of street and private parking space comfortably exceeds demand. But in higher-density urban areas that depend on street parking, the ability to fit more vehicles into the same length of street to satisfy previously unmet demand, or (depending on the urban design priorities) to reduce the percentage of the street that must be devoted to parking, is a very significant benefit. Space freed from parking might, for example, be devoted to trees, sidewalk cafés, and other amenities (figure 9.22).

There are similar opportunities in urban areas, shopping centers, campuses, and other contexts where off-street parking lots and structures are relied on. Either the capacities of parking facilities can be increased, or some of the space occupied by parking can be reallocated to other uses. The real estate value

released by this can be very significant and can help to cover the cost of transforming urban mobility systems, building the necessary new infrastructure, and enhancing public space. Figure 9.23, for example, illustrates a proposal for the city of Florence. Conventional automobile traffic is stopped at the ancient city walls, and underground parking structures are provided at the old city gates. Thus the car parking that currently blights most of the piazzas is eliminated, and these spaces are given back to the people for public and recreational use. Within the city walls, there is a mobility-on-demand system utilizing several vehicle types—USVs, battery-electric motor scooters, and bicycles. These vehicles not only provide clean, silent personal mobility within the historic city, but also have sufficient range to connect the historic city to the surrounding metropolitan area where employment opportunities, shopping centers, and other amenities not offered by the historic center are to be found.

Reducing the ratio of parking space to human activity space also changes the texture of urban areas, eliminating many of the "dead" spots created by parked cars and providing greater continuity of activity. Combined with reduced traffic noise, reduced local air pollution, and reduced danger from automobiles, this creates opportunities to return streets to their traditional urban role. They can become attractive, lively, pedestrian public spaces once again.

Finally, mobility-on-demand systems promise to transform the way we enter buildings and to remake the relationship of buildings and streets.

Typical Manhattan block (86 parking spaces)

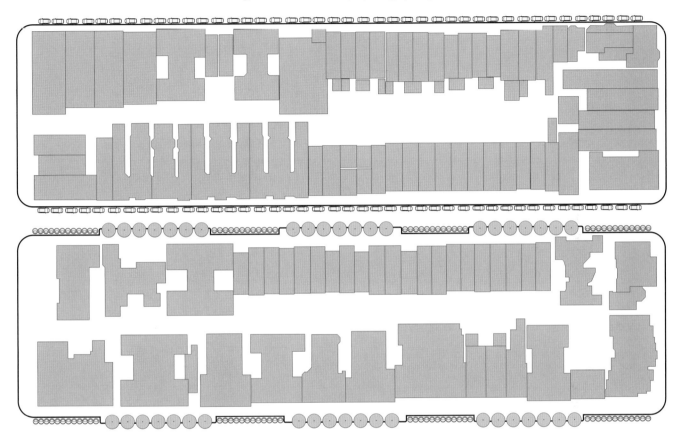

CityCar parking with 8 stations with 12 cars each (96 cars)

Figure 9.22
Improvements in street amenity enabled by parking
space reduction.

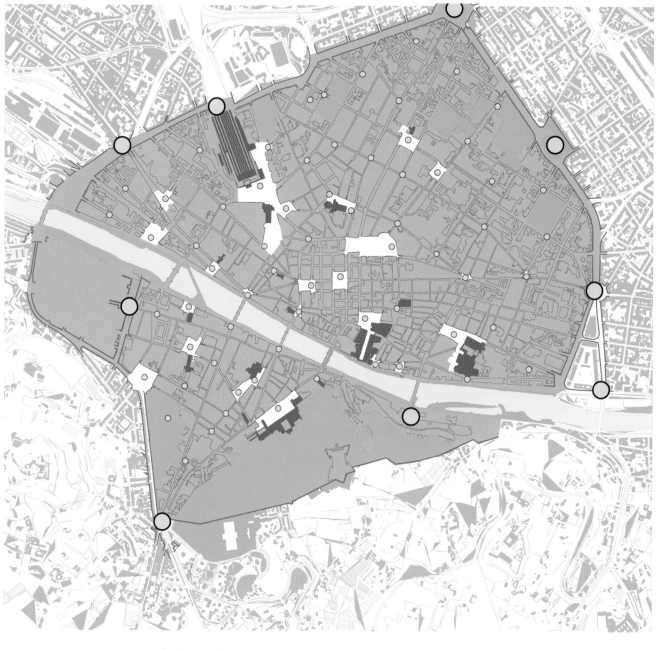

Where mobility-on-demand systems provide urban mobility, there is rarely a need to enter parking structures. Vehicles can be picked up and dropped off at the front doors of buildings.

Summary: Livable, Sustainable Cities

Throughout history, cities and the vehicles that provide mobility within them have coevolved. Each has adapted to the forms and requirements of the other. Reinventing the automobile now opens up the opportunity for cities to evolve in desirable new directions. And this new evolutionary path is essential for long-term livability and sustainability.

Many of the world's older cities, such as the camel-based cities of the medieval Middle East and North Africa, made little or no use of wheeled transportation.[3] Within these cities was typically a good match between the stability and agility of two-legged and four-legged animals and an urban pattern of narrow, winding streets, steps, and inclines. Since masses and speeds were very low there were few traffic safety concerns. Roman roads, on the other hand, provided generous, hard, flat surfaces that beautifully matched the capabilities of animal-powered wheeled vehicles. (And you could get run down by a speeding chariot.)

In general, where stability and agility are primary concerns, legs will have an advantage, but where mechanical efficiency and speed are more important, and there is a large investment in flat surfaces, wheels will begin to dominate.

When, at the beginning of the industrial revolution, steam power began to replace animal power in wheeled vehicles, steam trains motivated investment in hard, flat steel rails for them to run on, while the development of increasingly extensive rail networks increased the usefulness of locomotives and carriages and the demand for them. This, in turn, encouraged a pattern of concentrating population and facilities around rail stops and hubs.

The sprung wheels of early automobiles enabled them to navigate the roads and streets—frequently rough and unpaved—that had been created for pedestrians, horses, and low-speed animal-powered vehicles. But as automobiles became more powerful and faster, they generated a demand for better roads that would enable safer and more efficient operation, while the growing ubiquity of such roads encouraged further automobile use.[4] The long-term result has been the evolution of cities that are dominated by street and road systems that are closely adapted to the requirements of automobile traffic.

Figure 9.23
A proposal for use of USVs, battery-electric motor scooters, and bicycles to provide a mobility-on-demand system in the historic center of Florence.

Simultaneously, the provision of parking space has emerged as a crucial urban design issue. Streets are often choked with parked cars, and off-street parking lots and structures consume huge quantities of urban space that could be put to other uses.

Furthermore, the need for automobile carriageway and parking space has encouraged the widespread paving of urban land, reducing opportunities for greenery and natural drainage and exacerbating urban heat island effects. Streets have become dangerous, noisy sites of reduced air quality, with the consequence that their traditional role as public social space has diminished, and adjacent buildings, which might otherwise take advantage of natural ventilation, often have to be sealed up and air conditioned.

At a larger scale, the widespread use of gasoline-powered automobiles has encouraged inefficient and inconvenient land use patterns. Many cities have developed low-density urban fringes that are highly inefficient in their use of land and energy, are economically vulnerable, and offer few social and cultural opportunities.

These conditions are all well known and often thought to be inevitable—or, at least, inseparable from personal urban mobility. And their entrenched, ubiquitous character generates skepticism about the possibility of fundamental change. But this ignores the lessons of history and the principles of urban transformation. USVs have some compelling advantages. Like any invasive species, they will first take root, in competition with traditional automobiles, in urban contexts that are particularly congenial to them—where those advantages are most telling. This will initiate the familiar process of coevolving, mutually adapting vehicle and urban forms. Over time, as this process gathers momentum, we will see the increasing emergence of cities that provide high levels of personal mobility but are safer, quieter, cleaner, more livable, more energy efficient, and more sustainable than those of today. Cities that don't adapt fast enough will find themselves at an increasing competitive disadvantage with those that do.

Can you imagine the crowded streets of Manhattan filled with clean, quiet USVs? Can you imagine them rich with greenery and pedestrian amenities, and surrounded by naturally ventilated buildings? Perhaps it's difficult right now. But remember, those streets were once filled with horses.

Realizing the Vision

We have proposed ways to move freely, safely, and conveniently about our cities and towns in efficient, clean vehicles that are fashioned to our tastes, are fun to drive and ride in, and keep us linked seamlessly to our family, social, and business networks. Figure 10.1 summarizes the essence of our proposal for reinventing the automobile. Three major trends—growing urbanization, the electrification of energy and mobility systems, and the ongoing digital revolution in all aspects of telecommunication and information processing—are converging to create a space for radically reinventing personal urban mobility. Specifically, electrification and ubiquitous connectivity jointly enable USVs—Ultra Small Vehicles that are particularly well adapted to the needs and constraints of urban life. At urban scales, electrification combines with urban systems management and optimization to enable the synergistic integration of these electric vehicles with charging infrastructure and smart electric grids. Similarly, these management and optimization techniques combine with connectivity to enable large-scale, real-time collection of traffic and other data and the transmission of price signals back to automobiles. This makes possible efficient allocation of road space, parking space, and vehicles to meet dynamically varying mobility needs. All of these components come together in a comprehensive vision for reinvention not only of vehicles, but also of the personal urban mobility systems they support and the urban patterns to which they are wedded.

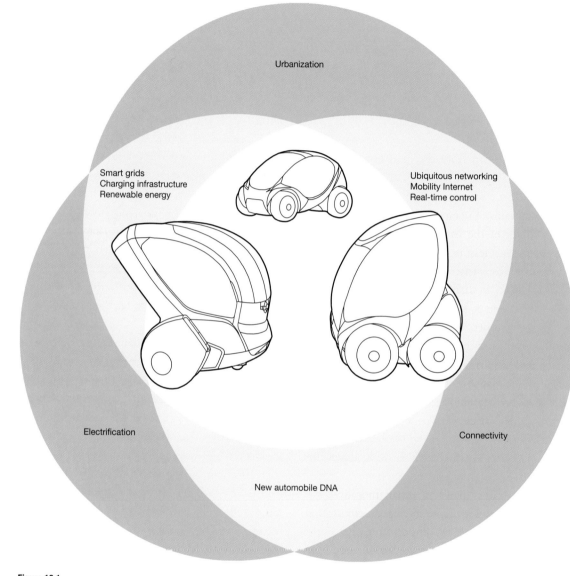

Urbanization

Smart grids
Charging infrastructure
Renewable energy

Ubiquitous networking
Mobility Internet
Real-time control

Electrification

Connectivity

New automobile DNA

Figure 10.1

The intersection of growing urbanization, electrification
of energy and mobility systems, and the ongoing digital
revolution creates a space for radically reinventing
personal urban mobility.

We believe this is technologically feasible and, better yet, we believe it can be done for less than one-quarter the cost of owning and operating one of today's typical automobiles.

But how do we make this vision real?

The Lessons of the Internet

First we must learn from the invention, development, and widespread acceptance of network-based systems with the scale and scope of personal urban mobility systems. The rapid adoption and continuing evolution of the Internet, in particular, offers many important lessons.

First, it is clear that large-scale networks depend on inexpensive, desirable devices to provide widespread access and to serve as network nodes. In the case of the Internet, the appearance of inexpensive personal computers in the early 1980s, followed by laptops and then smart phones, spurred more than two decades of rapid growth, eventually making Internet connectivity ubiquitous. Growth in the market for these devices, in turn, enabled investment in technological innovation and the achievement of economies of scale, which further drove down the cost of access.

In the case of personal urban mobility networks, the access devices are intelligent, connected, electric-drive automobiles—the counterparts of personal computers. Like PCs compared to mainframe computers, they are much smaller, lighter, and less expensive than their predecessors. Furthermore, they are essentially consumer electronics devices—

networked computers on wheels—relying for their functionality far less on mechanical and structural components, and far more on electronics and software, than traditional automobiles. Thus their costs can continually be driven down, and their performance improved, in the same ways as the costs and performances of computers.

Second, these devices should provide access to appealing new services. Personal computers, linked to the Internet, provide access to e-mail, Web pages, electronic commerce, and social networks. Apple's iPods provide access to iTunes and transformed music distribution. Amazon's Kindle reader provides access to wirelessly downloaded books. The more devices there are in use, the larger the market is for these services, and simultaneously, the more services there are available, the greater the motivation to acquire the devices.

Similarly, in personal urban mobility networks, intelligent, connected automobiles do not just provide transportation. They also provide enhanced awareness of other vehicles and of the surrounding environment, yielding safer and more efficient travel. And they provide awareness of current prices for electricity, road space, parking space, and in some contexts shared vehicle use, and, through intelligent response to price signals, the ability to use those resources optimally to achieve mobility goals.

Third, open standards such as TCP/IP (Transmission Control Protocol/Internet Protocol) and HTTP (Hypertext Transfer Protocol) enabled the explosive growth of the Internet and the Web. These are not closed, proprietary systems constrained by the strategy and investment capability of one organization. Anyone who complies with the relevant standards can become part of them and can interoperate with other parts. Thus the Internet and the Web mostly grew not in a top-down, planned fashion, but as the result of many dispersed and largely uncoordinated efforts. They snowballed as individuals and smaller networks linked to them and began to interoperate.

Growing personal urban mobility systems will also require: widely accepted, open standards; encouragement of bottom-up development and deployment efforts; avoidance of closed proprietary systems; and interoperability of vehicles produced by different companies and systems run by different operators. In particular, effective standards will be needed for telecommunication, connection to charging infrastructure and electric grids, and (much as with USB devices plugging into USB ports of computers) hardware interfaces among crucial components and subsystems.

Finally, the growth of the Internet has been accomplished through a complex combination of public and private initiatives and investments. The initial research and development, and the first phase of growth, resulted from large-scale government funding from ARPA (Advanced Research Projects Agency) and NSF (National Science Foundation). When the system had reached the necessary scale, it began to attract investments from infrastructure, equipment, and software companies, and it spawned many startups seeking to take advantage of the business opportunities that it offered.

Growth to pervasiveness has largely been driven by positive network externalities—the increasing utility of a network with every new node. If you have e-mail and only one other person has it, it is only marginally useful; if ten other people have it, it is far more useful; and if just about everyone has it, it is indispensable. The larger the network becomes, the more attractive it is to invest in adding to it, and even more so for the next addition. With open standards and interoperability, it is much easier to exploit this snowballing effect than within closed systems.

It is likely to be the same with personal urban mobility networks. There will be a need for some large-scale initial investments—which could come from public or private sources, or a combination—to initiate the process. From there onward, the more pervasive intelligent, connected electric-drive automobiles there are on the road, the greater the opportunities for cooperation in the interests of safety and efficiency. The more vehicles there are creating demand, the more attractive it is to invest in the charging infrastructure. The more widespread the charging infrastructure, the more attractive it is to acquire or use a vehicle. The more vehicles there are connected to smart grids, the stronger the synergies between vehicles and smart grids. Overall, the larger the network of vehicles and parking/charging stations, the more attractive it is to add to it. The goal, in initial investments and deployments, should be to start the necessary snowballing process.

The Internet and the World Wide Web began at modest scale but contained the seeds of explosive growth in these ways, and eventually became pervasive. Personal urban mobility systems have similar properties, and can grow in the same ways.

The Challenge of Large-Scale Codependencies

However, in applying the lessons of the Internet to personal urban mobility systems, we must recognize and develop effective strategies for overcoming the enormous inertia in today's automobile transportation system. This inertia results from scale and codependency.

As we discussed earlier, in the United States alone there are 250 million cars and trucks, 203 million licensed drivers, 4 million miles of roads, and 170,000 service stations. In addition, 14 million people make their living producing, selling, and servicing automobiles, producing and distributing fuel, and building, maintaining, and policing roads. Transforming a system of this scale is a daunting undertaking to say the least.

System codependencies add to the challenge. Automobiles depend on roads and conveniently available and affordable fuel. Roads are typically financed by fuel taxes. More cars mean more driving, more fuel consumption, and more tax revenue, leading to more and better roads. And a better road network means that cars provide more mobility value, which leads to even more cars. Furthermore, road designs depend on vehicle attributes, vehicles are designed to meet roadway standards, and fuel specifications must align with engine requirements, which are driven by emissions and fuel economy regulations. Finally, because today's vehicles are

human controlled, changing how we drive is a big deal, requiring hundreds of millions of drivers to be retrained. Clearly—and perhaps discouragingly, upon first consideration—everything in the system is strongly bound to everything else.

Given the inertia in today's automobile transportation system, it is not surprising that automobiles and their codependent roadway and energy infrastructures have improved gradually over time through incremental, evolutionary advances rather than through radical transformations. But, as is characteristic of highly evolved and integrated systems, they have reached a point where further evolution will have increasingly marginal effects.

This is similar to the situation of the communications industries in the 1960s, 1970s, and 1980s, when the Internet and the PC burst on the scene. The telephone, radio, television, newspaper publishing, book publishing, and mainframe computer industries were all large, highly evolved, in many cases heavily regulated, and enmeshed in complex webs of codependencies. Then their technological and business models were challenged by a series of disruptive innovations—packet switching and digital networking, inexpensive semiconductors, personal computers, multimedia over IP, e-mail, Web sites and browsers, search engines, and many more. From these disruptions, radically different products and systems, business models, and businesses emerged. The result was vastly improved communication, some highly profitable new businesses, and an era of extraordinary economic growth. The ideas that we have advanced in this book provide the starting points for a similarly creative disruption of personal urban mobility.

Working a Wicked Problem

The large-scale transformation of personal urban mobility systems through technological, design, and business innovation is a classic example of what Horst Rittel and Melvin Webber—writing more than three decades ago—defined as a "wicked problem."[1] It involves a system of highly interdependent systems, with the property that actions taken to improve one aspect of the system may produce unexpected reactions and unwelcome side effects. It is complex, ambiguous, and defies any straightforward progression from defining goals, through designing and engineering solutions, to manufacture of products and integration and deployment of systems. It is not like getting a man to the moon. It requires creative speculation about possibilities, ongoing critical discussion of principles and options, engagement of stakeholders with differing and perhaps conflicting interests, building consensus and coalitions of interest, and responding flexibly to the unexpected twists and turns that emerge along the way to a solution.

Urban designers and planners are very familiar with wicked problems and strategies for dealing with them. Product designers and engineers are, perhaps, less so. But prominent business writers have increasingly engaged with the idea—particularly when pursuing transformational change—and have some useful suggestions to offer. We note a few of these writers and their ideas here:

- **Russell Ackoff** "Idealized design" recommends starting with the desired end in mind (an ideal solution) and then working backward to where you are today to develop a roadmap to your destination. It focuses on driving fundamental, transformative change. Idealized design helps eliminate obstacles to change and "dissolves" problems by designing systems that prevent them from arising in the first place.

- **Roger Martin** The "business of design" approach suggests that traditional problem solvers tend to follow a linear model based on gathering data to understand the problem, analyzing these data to formulate a solution, and then implementing this solution to solve the problem. In contrast, designers tend to follow a less linear model. They begin by trying to understand the problem, then quickly move to formulating potential solutions, and then jump back to refining their understanding of the problem. Designers know that understanding of a problem can only arise from creating possible solutions and that this understanding continues to evolve hand in hand with increasingly better solutions.

- **Anita McGahan** Through numerous case studies, McGahan demonstrates that the defining issues of this century (including starvation, disease, terrorism, climate change, sustainable economic growth, and sustainable mobility) tend to be wicked problems with solutions that simultaneously require technical innovation (in products, processes, and services), organizational innovation (including new ways of organizing collaborations among businesses, government, nongovernmental organizations, and international multilateral agencies), and restructuring of social contracts.

- **Clayton Christensen** The "innovator's dilemma" proposes principles of disruptive innovation that can effectively counteract the inertia created by existing product and service solutions to the "jobs" that consumers want done. Existing solutions are often overspecified, providing the opportunity for innovators to attack "under the radar screen" with simpler, lower-cost designs that more optimally meet customers' needs. The Ultra Small Vehicles introduced in chapter 4 fit this description. Christensen also emphasizes the importance of focusing on compelling value propositions, market "footholds," and rapid learning cycles to realize transformational change.

- **Malcolm Gladwell** The "tipping point" approach shows how an idea, trend, or social behavior can cross a threshold, tip a system, and then spread extremely rapidly. Tipping is enabled by the fact that human beings are profoundly social beings, influenced by and influencing other human beings. Realization of transformational change in the face of enormous inertia requires the identification and pursuit of tipping points.

We can begin to move forward by combining some of the lessons of Ackoff, Martin, McGahan, Christensen, and Gladwell—embarking on a journey to where we need to go, even if we end up somewhere different from where we were initially headed. Disruptive innovation focused on wicked problems can

trump inertia if there is sufficient will for it, and if we improvise along a roadmap created with the end in mind, while being driven first to reach tipping points efficiently, quickly, and, at times, stealthily.

From an idealized design perspective, the end we have in mind is to enable people to move about and interact with greater freedom and more enjoyment than we realize today without any negative side effects (that is, without air pollution, nonrenewable energy consumption, deaths, injuries and property damage, wasted time, and inequality of access). The outline of our ideal solution described in chapter 9 was based on integrating the four building blocks introduced in chapters 2–8. To realize this solution, Ackoff would advise working backward from this outline to create a roadmap of planned actions that will mature, synchronize, and converge the required enablers.

From an innovator's dilemma perspective, several demonstration and market test "footholds" already exist for our required enablers:

- Battery-electric, extended-range electric, and fuel cell electric vehicles are being developed and marketed.
- Wheel motors for propulsion, braking, and steering have been demonstrated and used by numerous automotive and specialty companies (Chevrolet Sequel, CityCar, Segway PT, and so on).
- Telematics services are becoming widespread. (OnStar, for example, has around 6 million customers.)
- Information-rich digital maps and GPS-based navigation systems are widely available and are continuing to improve.

- Adaptive cruise control and lane-departure warning and lane keeping have been or are being commercialized in passenger vehicles.
- GM has demonstrated vehicle-to-vehicle and vehicle-to-pedestrian/cyclist communications and has patented a PDA-sized transponder to enable this capability.
- Autonomous driving has successfully been demonstrated in the DARPA Urban Challenge, and Abu Dhabi is pioneering Masdar Personal Rapid Transit (PRT) podcars.
- Road pricing is being used to manage congestion and reduce automobile usage in London, Singapore, and elsewhere.
- Car-free zones have been established in New York City and Amsterdam.
- Bicycle sharing and car sharing (ZipCar and I-GO) already exist.
- One-way rental, mobility-on-demand systems have been demonstrated, at a large scale, in bicycle-sharing schemes.

The key now is to learn quickly from these footholds with an intense focus on customers (that is, individuals seeking personal mobility and society seeking to eliminate negative externalities) and integrate these initiatives into an idealized design roadmap.

From a tipping point perspective, we must define, continuously modify (based on what we learn as we go), and steer our roadmap so that the enablers of our vision reach market tipping points (either individually or in subsets) quickly and efficiently. This approach is consistent with the way that designers create initial solutions that redefine problems leading to ever-better solutions and understanding of the problem. This design process initially must be focused on attaining tipping points to kick-start a self-sustaining growth dynamic leading to widespread use.

Finally, to address the "wickedness" of the sustainable mobility challenge, we must pursue organizational innovation and new social contracts. The codependencies underlying today's personal mobility solutions make it clear that no one company, industry, or government working alone can bring about transformational change. We must collaborate and find new ways to partner together to realize our ideal solution.

Summary: The Essential Steps

Here in broad outline is what has to happen to turn the idea of sustainable personal mobility into reality:

1. *We must build as broad a consensus as possible, among those concerned with personal urban mobility, concerning the elements and priorities of the long-term vision.* The stakeholders include automobile companies and related enterprises, transportation operators, information technology companies, electric utilities and energy companies, real estate developers, urban planners and designers, regulation and policy institutions, and political leaders at multiple levels. The consensus among them cannot and will not be complete, but if there is sufficient buy-in it will provide a workable foundation for moving forward. To achieve this we must recognize the differing and

sometimes conflicting interests of the various stakeholders, take account of the complex codependencies that are involved, find creative ways to align interests and goals, and discourage narrowly framed suboptimization of the pieces. This will provide the necessary common understanding, motivation, and collective will.

2. *We must develop a comprehensive transformation roadmap for achieving the vision.* This will require leadership to bring the key stakeholders to the table, forums to develop the roadmap details, and organizational innovation—particularly in the area of public–private partnership. It must define ways to share the private and public risks and rewards inherent in transformational change.

3. *The roadmap must link together existing and future "footholds" to test and improve the enablers of our ideal solution with an eye to reaching market tipping points fast and at minimal cost.* Markets must ultimately drive widespread, high-volume demand for new forms of personal mobility. Consumers will not purchase in high volume if prices exceed value. Companies cannot afford to sell at high volume if cost exceeds price. Governments cannot afford to subsidize high volumes. The transformation roadmap must comprehend these realities and accept that technology matures to its ultimate value and cost potential over a series of generational improvements. Each generation provides essential consumer, product, and supplier learning cycles as the system progresses toward the sought-after tipping points. This transition must be managed, and all stake-

holders must have the mutual trust and courage to work together on achieving it.

4. *The roadmap must include a portfolio of alternative ways to reach the desired end state.* It is clearly unwise to bet on just one "silver bullet" technology and route. A differentiated portfolio balances risk and return. We believe the vision for sustainable personal mobility proposed in this book is very compelling and promising, but, like most visions, it is fraught with uncertainties. We should develop multiple options and refine them as we learn. The stakes are too high to overplay one hand.

5. *We must develop enabling standards and pursue ways of harnessing positive network externalities.* Enabling standards, such as TCP/IP and HTTP, opened the way to rapid growth of the Internet and the World Wide Web, and they can do the same for personal urban mobility systems. Proprietary standards and closed systems may confer business advantages in the short term, but in the long term these are enormously outweighed by the advantages of interoperability among subsystems. Networks grow rapidly when new pieces can be added, without difficulty, by anyone, and when each addition increases the value of the network to all those who use it.

6. *We must engage imagination and desire.* No new system of this sort can succeed unless its potential users understand it and want it. And it is difficult to understand—much less desire—abstractions and technicalities. We will need diverse, imaginative design explorations, compelling narratives, and convincing prototypes and pilot projects.

Sustainable personal mobility is simultaneously an essential need, an extraordinary opportunity, and a wicked problem. As Roger Martin has eloquently put it: "Wicked problems call for us to harness all of the creativity and knowledge at our disposal. By working to enable shared understanding and commitment, we have the collective power to shape our world for the better. Whether we choose to fight one another or work together to confront opportunities and threats, our fate and our common wealth are in our hands."[2]

We have in our hands the opportunity to reinvent the automobile, create more functional and efficient personal urban mobility systems, redirect the evolution of our cities along a more sustainable path, and create clean, green economic growth. This opportunity entails redefining the automobile's DNA, creating the Mobility Internet, pursuing a clean, smart energy supply, and developing dynamically priced, electronically managed markets for electricity, road space, parking space, and vehicles.

Notes

Chapter One

1. Thomas Brinkhoff, "The Principal Agglomerations of the World," http://www.citypopulation.de/world/Agglomerations.html.

2. U.S. Department of Transportation, Bureau of Transportation Statistics, *National Transportation Statistics 2008*, http://www.bts.gov/publications/national_transportation_statistics/2008/index.html; US Department of Energy, Energy Information Administration, http://www.eia.doe.gov/.

3. General Motors Global Market and Industry Analysis.

4. International Energy Agency, http://www.iea.org/; World Energy Outlook 2008 http://www.worldenergyoutlook.org/; GM Global Energy Systems Intelligence Center.

5. World Health Organization/World Bank, *The World Report on Road Traffic Injury Prevention* (Geneva: World Health Organization, 2004).

6. Jeffrey R. Kenworthy and Felix B. Laube, *An International Sourcebook of Automobile Dependence in Cities: 1960–1990* (Boulder: University Press of Colorado, 2000).

7. The Internet of things is based on the idea that every physical object can be electronically tagged, sensed, given an IP address, and connected to the Internet. For a cogent introduction see Sean Dodson, "The Net Shapes Up to Get Physical," *Guardian*, October 16, 2008.

Chapter Two

1. The best-known steam car in the United States was the Stanley Steamer, manufactured from 1896 to 1924.

2. Electric vehicles were sold in the United States in the early twentieth century by companies such as Detroit Electric, Edison, and Studebaker.

3. Charles and Frank Duryea initiated America's automobile industry in 1893 by "mass-producing" thirteen gasoline-fueled cars.

4. The electric self-starter debuted on the 1912 Cadillac.

5. Olds patented the assembly line concept in 1901. Henry Ford greatly extended its implementation in 1913 with the moving assembly line.

6. Source: Edmunds Inc. True Cost to Own (TCO) pricing tool includes depreciation, interest on your loan, taxes and fees, insurance premiums, fuel costs, maintenance, and repairs over five years, for 15,000 miles per year. See http://www.edmunds.com/apps/cto/CTOintroController.

Chapter Three

1. Jacob Eriksonn, Hari Balakrishnan, and Samuel Madden, "Cabernet: Vehicular Content Delivery Using WiFi," *14th ACM MOBICOM*, San Francisco, September 2008.

2. For extensive discussion of this issue see Donald Shoup, *The High Cost of Free Parking* (Chicago: American Planning Association, 2005).

Chapter Four

1. Julia King, *The King Review of Low-Carbon Cars, Part 1: The Potential for CO_2 Reduction* (London: HMSO, 2007).

2. P.U.M.A. was a GM-funded development program with Segway. See http://www.segway.com/puma/.

3. This was developed by Franco Vairani and documented in his 2009 MIT doctoral thesis.

4. "Your Driving Costs," http://www.AAA.com/public_affairs.

5. Colliers International 2008 Parking Rate Survey, http://www.colliersmn.com.

Chapter Five

1. Jago Dodson and Neil Sipe, *Shocking the Suburbs: Oil Vulnerability in the Australian City* (Sydney: University of New South Wales Press, 2008).

Chapter Six

1. It is easy to forget that gasoline filling stations were at one time themselves an innovation. John D. Rockefeller made much of his fortune from the prescient business move of creating a ubiquitous system of filling stations for proliferating automobiles. And this reinforced the advantage of early gasoline-powered cars over competing steam and electric cars.

2. Thomas P. Hughes, *Networks of Power: Electrification in Western Society, 1880–1930* (Baltimore: Johns Hopkins University Press, 1983); David E. Nye, *Electrifying America: Social Meanings of a New Technology* (Cambridge, Mass.: MIT Press, 1990).

3. Ryan Randazzo, "Valley Plans to Set Up Charging Stations for Electric Cars," *Arizona Republic*, April 16, 2009.

4. Jonathan Soble, "Japan Fuels Electric Car Revolution," *Financial Times*, August 25, 2008.

5. Despite this disadvantage, some early electric buses depended on battery swapping, and Better Place electric cars have recently revived it. See http://www.betterplace.com.

6. Byoungwoo Kang and Gerbrand Ceder, "Battery Materials for Ultrafast Charging and Discharging," *Nature* 458 (March 12, 2009): 190–193.

7. The APS system was developed by Innorail, part of Alstom.

8. Aristeidis Karalis, J. D. Joannopoulos, and Marin Soljačić, "Efficient Wireless Non-Radiative Mid-Range Energy Transfer," *Annals of Physics* 323 (2008): 34–48; André Kurs, Aristeidis Karalis, Robert Moffatt, J. D. Joannopoulos, Peter Fisher, and Marin Soljačić, "Wireless Power Transfer via Strongly Coupled Magnetic Resonances," *Science* 317 (July 6, 2007): 83–86.

Chapter Seven

1. Cogeneration is the use of a single primary heat source (usually coal, oil, or natural gas) to produce electricity and heat simultaneously. If cogeneration is done within or near buildings, the heat can be used for space heating and hot water. Typically, cogeneration saves between 10 percent and 30 percent of the fuel that would be used for separate production of electricity and heat. See Neil Petchers, *Combined Heating, Cooling, and Power Handbook: Technologies and Applications: An Integrated Approach to Energy Resource Optimization* (Lilburn, Georgia: Fairmont Press, 2002).

2. The concept of smart sustainability was introduced in William J. Mitchell, "Lean and Green," chapter 10 of *E-topia* (Cambridge, Mass.: MIT Press, 1999), pp. 146–155.

Chapter Eight

1. One technique for accomplishing this is described in Raluca Ada Popa, Hari Balakrishnan, and Andrew J. Blumberg, "VPriv: Protecting Privacy in Location-Based Vehicular Services," *18th USENIX Security Symposium*, Montreal, August 2009.

Chapter Nine

1. Keith Bradsher, "China Vies to Be World's Leader in Electric Cars," *New York Times*, April 2, 2009.

2. Don Shoup, *The High Cost of Free Parking* (Chicago: American Planning Association, 2005).

3. Richard W. Bulliet, *The Camel and the Wheel* (New York: Columbia University Press, 1975).

4. Peter D. Norton, *Fighting Traffic: The Dawn of the Motor Age in the American City* (Cambridge, Mass.: MIT Press, 2008).

Chapter Ten

1. Horst Rittel and Melvin Webber, "Dilemmas of a General Theory of Planning," *Policy Sciences* 4 (1973): 155–169.

2. Roger Martin, "Introduction," *Rotman: Magazine of the Rotman School of Management, University of Toronto*, Special Issue on Wicked Problems, winter 2009, p. 3.

Acknowledgments

The authors are particularly grateful to Ryan Chin for his extremely effective management of the CityCar project in the Smart Cities group at the MIT Media Laboratory, and for his supervision of the production of the illustrations for this book.

We are also particularly grateful to William Lark, Jr., for serving as lead designer on the version of the Smart Cities City-Car that is presented throughout this book.

Many of the ideas and designs that are discussed in this book emerged from General Motors–sponsored research in the Smart Cities group at the Media Laboratory. Over several years, this involved many graduate research assistants, undergraduates participating in MIT's UROP (Undergraduate Research Opportunities Program), participants in associated courses, faculty members and researchers from elsewhere at MIT, and visitors. The following is a list of contributors. Our sincere apologies if we have inadvertently omitted anyone.

Smart Cities

Ryan Chin
Chih-Chao Chuang
Wayne Higgins
Mitchell Joachim
Axel Kilian
Patrik Künzler
William Lark, Jr.
Philip Liang
Michael Chia-Liang Lin
Neri Oxman
Dimitris Papanikolaou
Arthur Petron
Raul-David "Retro" Poblano
Somnath Ray
Peter Schmitt
Susanne Seitinger
Andres Sevtsuk
Franco Vairani

General Motors

David Cameron
Wayne Cherry
Gary Cowger
Roy Mathieu
Mike Peterson

MIT

Julius Akinyemi
Stephen Benton
Mike Bove
Federico Casalegno
John Difrancesco
Ed Fredkin
Dennis Frenchmen
Ralph Gakenheimer
Neil Gershenfeld
Daniel Greenwood
John Hansman
John Heywood
Kent Larson
John Maeda
Betty Lou McClanahan
Marvin Minsky
Frank Moss
Joseph Paradiso
Donald Sadoway
Michael Schrage
Glen Urban
Chris Zegras

Illustration Production

Ryan Chin
Chih-Chao Chuang
William Lark, Jr.
Dimitris Papanikolaou
Ruifeng Tian

UROP

Abdulaziz Albahar
Josh Bails
Pablo Bello
Tom Brown
Nathaniel Forbes
Charles Guan
Robert Han
Itaru Hiromi
Brian Hong
Deke Hu
Cathie Kim
Pall Kornmayer
Andrew Leone
Jeanna Liu
Daniel Lopuch
Jason Martinez
Nicholas Pennycooke
Marcus Parton
Adam Paxson
Brad Schiller
Laura Schuhrke
Patrick Shannon
Laurie Stach
Edgar Torres
John Williams
Alison Wong
Allen Yin

Outside of MIT

Guilia Biamino
Bono
Topper Carew
Cristiano Ceccato
Edge
Alex Fiechter
Frank O. Gehry
James Gips
James Glymph
Ralph Hulseman
Mel King
Roberto Montanari
Laura Noren
Don Rembowski
Moshe Safdie
Dennis Sheldon
Dan Williams

MIT Course Participants

Claire Abrahamse
Anas Alfaris
Robyn Allen
Laura Aust
Marcel Botha
Louis Basel
Mike Beaser
Luis Berrios-Negron
Marcel Botha
Christine Canabou
Brian Chan
Darren Chang
Young Joong Chang
Mark Cote
Omari Akil Davis
Sloan Dawson
Chad Dyner
Alexis Fiechter
Kweku Fleming
Victor Gane
Enrique Garcia
Lorene Gates-Spears
David Jason Gerber
Jonathan Gips
Joshua Goldwitz
Yehuda Greenfield
Onur Yuce Gun
Ziga Ivanic
Carrie Huang
Huiying Ke
Jae Kyung Kim
Cha-Ly Koh
Sotirios Kotsopoulos
Inna Koyrakh
Ashwani Kumar
Shelley Lau
August Liau
Johan Lofstrom
Daniel Lopuch
Yanni Loukissas

Anmol Madan
Arthur Mak
Daniel McLaughlin
Esmeralda Megally
Bryan Morrissey
Olumuyiwa Oni
Christine Outram
Michael Pierce
Pam Rae Pitchot
Timocin Pervane
Randolph Punter
Christianna Raber
Kalin A. Rahnev
Daniel Rosenberg
Yang Ruan
Costantino Sambuy
Jota Samper
David Spectre
Kate Tan
Christopher Taylor
James Chao-Ming Teng
Bo Stjerne Thomsen
Ruifeng Tian
Zenovia Toloudi
Matt Trimble
Isabella Tsao
Maya Turre
Lars Wagner
Conor Walsh
Eric Weber
Andrew John Wit
Chee Qi Xu
Polychronis Ypodimatopoulos
Giampaolo Zen

General Motors

In addition to the Smart Cities program, the authors would like to acknowledge the following persons who have contributed to the development at General Motors of vehicles that embody the reinvention of the automobile and personal urban mobility, vehicle electrification and connectivity, and autonomous vehicles. The authors are particularly grateful to Lerinda Frost for her long-standing support in communicating the "Reinvention of the Automobile."

Hong Bae
Chaminda Basnayake
Jon Bereisa
Bob Boniface
Matthias Bork
Nady Boules
Norm Brinkman
Tom Brown
Michelle Burrows
Al Butlin
Greg Cesiel
Adrian Chernoff
David Connor
Dan Demitrish
Tom DeMurry
Kevin Deng
Don Dibble
Scott Fosgard
Matt Fronk
Lerinda Frost
Susan Garavaglia
Omar Garcia
Chuck Green
Randy Greenwell
Don Grimm
Britta Gross
Rajiv Gupta
Rick Holman
Young Sun Kim
Hariharan Krishnan
Bob Lange

Jin-Woo Lee
Bakhtiar Litkouhi
Anthony Lo
Sean Lo
Tom Lobkovich
Joe Lograsso
Mike Losh
Kathy Marra
Lothar Matajcek
Renee McClelland
Byron McCormick
Dick McGinnis
Joe Mercurio
Mike Milani
Mike Miller
Pri Mudalige
Bernhard Mueller
Jim Nagashima
Sanjeev Naik
Jim Nickolaou
Alan Nicol
Markus Noll
Dan O'Connell
Massimo Osella
Larry Peruski
Mike Pevovar
Clay Phillips
Jeff Pleune
Patrick Popp
Tony Posawatz
Nick Pudar

David Rand

Tom Read

Dan Roesch

Varsha Sadekar

Jeremy Salinger

Cem Saraydar

Frank Saucedo

Christof Scherl

Mohsen Shabana

Robert Shafto

Angele Shaw

Todd Shupe

Jim Tarchinski

Alan Taub

Jussi Timonen

Hong Tran

David Tulauskas

Justin Twitchell

Ramasamy Uthurusamy

Robert Vitale

Terry Ward

Ed Welburn

Johan Willems

Chet Wisniewski

Ray Wokdacki

Jeff Wolak

David Young

Shuqing Zeng

Wende Zhang

Nick Zielinski

Bibliography

The literatures on urban form and function, transportation technologies and systems, automobiles, traffic, energy systems, and networks are vast. This bibliography provides a highly selective introduction, emphasizing works with particular relevance to the themes and arguments of this book.

Ackoff, Russell L., Jason Magidson, and Herbert J. Addison. *Idealized Design: How to Dissolve Tomorrow's Crisis Today*. Upper Saddle River, N.J.: Wharton School Publishing, 2006.

Adams, Ronald, and Terry Brewer. "A Plan for 21st Century Land Transport." *International Journal of Vehicle Design* 35 (1/2) (2004): 137–150.

Ausubel, Jesse H., Cesare Marchetti, and Perrin Meyer. "Toward Green Mobility: The Evolution of Transport." *European Review* 6 (2) (1998): 137–156.

Bacon, Edmund N. *The Design of Cities*. New York: Penguin, 1974.

Banister, David. *Unsustainable Transport: City Transport in the New Century*. New York: Routledge, 2005.

Barabási, Albert-László. *Linked: The New Science of Networks*. Cambridge, Mass.: Perseus Publishing, 2002.

Batty, Michael. *Cities and Complexity: Understanding Cities with Cellular Automata, Agent-Based Models, and Fractals*. Cambridge, Mass.: MIT Press, 2005.

Batty, Michael. "The Size, Scale, and Shape of Cities." *Science* 319 (5864) (2008): 769–771.

Ben-Joseph, Eran. *The Code of the City: Standards and the Hidden Language of Place Making*. Cambridge, Mass.: MIT Press, 2005.

Black, William A. *Transportation: A Geographical Analysis*. New York: The Guilford Press, 2003.

Blainey, Geoffrey. *The Tyranny of Distance: How Distance Shaped Australia's History*. Melbourne: Sun Books, 1966.

Boyle, Godfrey, ed. *Renewable Energy*. Oxford: Oxford University Press, 2004.

Boyle, Godfrey, ed. *Renewable Energy and the Grid: The Challenge of Variability*. London: Earthscan Publications, 2007.

Boyle, Godfrey, Bob Everett, and Janet Ramage. *Energy Systems and Sustainability*. Oxford: Oxford University Press, 2003.

Brandon, Ruth. *Auto Mobile: How the Car Changed Life*. London: Macmillan, 2002.

Bruegmann, Robert. *Sprawl: A Compact History*. Chicago: University of Chicago Press, 2005.

Bryant, Bob. "Actual Hands-off Steering and Other Wonders of the Modern World." *Public Roads* 61 (3) (1997).

Buchanan, Colin. *Traffic in Towns*. Harmondsworth: Penguin, 1963.

Bulliet, Richard W. *The Camel and the Wheel*. New York: Columbia University Press, 1975.

Burns, Lawrence D., Byron McCormick, and Christopher E. Borroni-Bird. "Vehicle of Change." *Scientific American* 287 (2002): 64–73.

Burns, Lee. *Busy Bodies: Why Our Time-Obsessed Society Keeps Us Running in Place*. New York: Norton, 1993.

Castells, Manuel. *The Internet Galaxy: Reflections on the Internet, Business, and Society*. New York: Oxford University Press, 2003.

Christenson, Clayton M. *The Innovator's Dilemma: When New Technologies Cause Great Firms to Fail*. Cambridge, Mass.: Harvard Business School Press, 1997.

DARPA. *DARPA Urban Challenge*. www.darpa.mil/GRAND-CHALLENGE/overview.asp (accessed January 13, 2009).

Dodson, Jago, and Neil Sipe. *Shocking the Suburbs: Oil Vulnerability in the Australian City*. Sydney: University of New South Wales Press, 2008.

Dodson, Sean. "The Net Shapes Up to Get Physical." *Guardian*, October 16, 2008.

Downs, Anthony. *Still Stuck in Traffic: Coping with Peak-Hour Traffic Congestion*. Washington, D.C.: Brookings Institution Press, 1992.

Droege, Peter. *The Renewable City: A Comprehensive Guide to an Urban Revolution*. Chichester: Wiley, 2006.

Droege, Peter. *Urban Energy Transition: From Fossil Fuels to Renewable Power*. Oxford: Elsevier Science, 2008.

Flink, James J. *The Automobile Age*. Cambridge, Mass.: MIT Press, 1990.

Frenchman, Dennis, Giandomenico Amendola, Anne Beamish, and William J. Mitchell. *Technological Imagination and the Historic City: Florence*. Naples: Liguori Editore, 2009.

Friedman, Thomas L. *Hot, Flat, and Crowded: Why We Need a Green Revolution—and How It Can Renew America*. New York: Farrar, Strauss, and Giroux, 2008.

Galvin, Robert, and Kurt Yeager. *Perfect Power: How the Microgrid Revolution Will Unleash Cleaner, Greener, and More Abundant Power*. New York: McGraw-Hill, 2009.

Gladwell, Malcolm. *The Tipping Point: How Little Things Can Make a Big Difference*. Boston: Little, Brown, 2002.

González, Marta C., César A. Hidalgo, and Albert-László Barabási. "Understanding Individual Human Mobility Patterns." *Nature* 453 (2008): 779–782.

Graham, Stephen, and Simon Marvin. *Splintering Urbanism: Networked Infrastructures, Technological Mobilities and the Urban Condition*. New York: Routledge, 2001.

Grava, Sigurd. *Urban Transportation Systems*. New York: McGraw-Hill, 2003.

Hall, Peter. *Cities in Civilization*. New York: Pantheon, 1998.

Henley, Simon. *The Architecture of Parking*. New York: Thames and Hudson, 2007.

Hoffert, Martin I., Ken Caldeira, Gregory Benford, David R. Criswell, Christopher Green, Howard Herzog, Atul K. Jain, Haroon S. Kheshgi, Klaus S. Lackner, John S. Lewis, H. Douglas Lightfoot, Wallace Manheimer, John C. Mankins, Michael E. Mauel, L. John Perkins, Michael E. Schlesinger, Tyler Volk, and Tom M. L. Wigley. "Advanced Technology Paths to Global Climate Stability: Energy for a Greenhouse Planet." *Science* 298 (5595) (2002): 981–987.

Holding, Erling. *Achieving Sustainable Mobility: Everyday and Leisure-Time Mobility in the EU*. Aldershot: Ashgate, 2007.

Hughes, Thomas P. *Networks of Power: Electrification in Western Society, 1880–1930*. Baltimore: Johns Hopkins University Press, 1983.

International Energy Agency (IEA). *World Energy Outlook 2008*. Paris: International Energy Agency, 2008.

Jacobs, Allan B. *Great Streets*. Cambridge, Mass.: MIT Press, 1995.

Jakle, John A., and Keith A. Sculle. *Lots of Parking: Land Use in a Car Culture*. Charlottesville: University of Virginia Press, 2005.

Kenworthy, Jeffrey R., Felix B. Laube, Tamim Raad, Chamlong Poboon, and Benedicto Guia. *An International Sourcebook of Automobile Dependence in Cities, 1960–1990*. Boulder: University Press of Colorado, 2000.

King, Julia. *The King Review of Low-Carbon Cars, Part 1: The Potential for CO_2 Reduction*. London: HMSO, 2007.

King, Julia. *The King Review of Low-Carbon Cars. Part 2: Recommendations for Action*. London: HMSO, 2008.

Kirsch, David A. *The Electric Vehicle and the Burden of History*. New Brunswick, N.J.: Rutgers University Press, 2000.

Klare, Michael T. *Rising Powers, Shrinking Planet: The New Geopolitics of Energy*. New York: Metropolitan Books, 2008.

Kostof, Spiro. *The City Shaped: Urban Patterns and Meanings Through History*. Boston: Bulfinch, 1993.

Ladd, Brian. *Autophobia: Love and Hate in the Automobile Age*. Chicago: University of Chicago Press, 2008.

Lovins, Amory, and D. R. Cramer. "Hypercars, Hydrogen, and the Automotive Transition." *International Journal of Vehicle Design* 35 (1/2) (2004): 50–85.

Lynch, Kevin. *Good City Form*. Cambridge, Mass.: MIT Press, 1984.

MacKay, David J. C. *Sustainable Energy—Without the Hot Air*. Cambridge: UIT, 2009.

Martin, Roger. "Introduction." *Rotman: Magazine of the Rotman School of Management, University of Toronto*, Special Issue on Wicked Problems (winter 2009).

Martin, Roger. *The Opposable Mind: How Successful Leaders Win Through Integrative Thinking*. Cambridge, Mass.: Harvard Business School Press, 2009.

Metz, David. *The Limits to Travel*. London: Earthscan, 2008.

Mitchell, William J. *City of Bits: Space, Place, and the Infobahn*. Cambridge, Mass.: MIT Press, 1994.

Mitchell, William J. *E-topia*. Cambridge, Mass.: MIT Press, 1999.

Mitchell, William J. *Me++: The Cyborg Self and the Networked City*. Cambridge, Mass.: MIT Press, 2004.

Mohan, Dinesh. *Mythologies, Metros, and Future Urban Transport*. TRIPP Report 08-01. New Delhi: Transportation Research and Injury Prevention Program, Indian Institute of Technology Delhi, 2008.

Mokhtarian, Patricia L., and Cynthia Chen. "TTB or not TTB, That Is the Question: A Review and Analysis of the Empirical Literature on Travel Time (and Money) Budgets." *Transportation Research Part A, Policy and Practice* 38 (9–10) (2004): 643–675.

Mom, Gijs. *The Electric Vehicle: Technology and Expectations in the Automobile Age*. Baltimore: Johns Hopkins University Press, 2004.

Moriarty, Patrick. "Environmental and Resource Constraints on Asian Urban Travel." *International Journal of Environment and Pollution* 30 (1) (2007): 8–26.

Mumford, Lewis. "The Highway and the City." In *The Highway and the City*, 244–256. New York: Mentor, 1964.

Mumford, Lewis. *The City in History: Its Origins, Its Transformations, and Its Prospects*. New York: Harvest Books, 1968.

National Research Council. *The Hydrogen Economy: Opportunities, Costs, Barriers, and R&D Needs*. Washington, D.C.: National Academies Press, 2004.

Newman, Jeffrey R., and Felix B. Laube. *An International Sourcebook of Automobile Dependence in Cities: 1960–1990*. Boulder: University Press of Colorado, 2000.

Norton, Peter D. *Fighting Traffic: The Dawn of the Motor Age in the American City*. Cambridge, Mass.: MIT Press, 2008.

Nye, David E. *Electrifying America: Social Meanings of a New Technology*. Cambridge, Mass.: MIT Press, 1990.

Pacala, Stephen, and Robert Socolow. "Stabilization Wedges: Solving the Climate Problem for the Next 50 Years with Current Technologies." *Science* 305 (August 13, 2004): 968–976.

Pentland, Alex (Sandy). *Honest Signals: How They Shape Our World*. Cambridge, Mass.: MIT Press, 2008.

Peters, Peter Frank. *Time Innovation and Mobilities*. London: Routledge, 2006.

Popa, Raluca Ada, Hari Balakrishnan, and Andrew J. Blumberg. "VPriv: Protecting Privacy in Location-Based Vehicle Services." Presented at 18th USENIX Security Symposium. Montreal, August 2009.

Puentes, Robert, and Adie Tomer. *The Road . . . Less Traveled: An Analysis of Vehicle Miles Traveled Trends in the US*. Metropolitan Infrastructure Series, Metropolitan Policy Program at Brookings. Washington, D.C.: Brookings Institution, 2008.

Resnick, Mitchel. *Turtles, Termites, and Traffic Jams: Explorations in Massively Parallel Microworlds*. Cambridge, Mass.: MIT Press, 1997.

Rodrigue, Jean-Paul, Claude Comtois, and Brian Slack. *The Geography of Transport Systems*. London: Routledge, 2006.

Safdie, Moshe, with Wendy Kohn. *The City After the Automobile: An Architect's Vision*. Boulder, Colo.: Westview Press, 1998.

Schäfer, Andreas, John B. Heywood, Henry D. Jacoby, and Ian A. Waitz. 2008. *Transportation in a Climate-Constrained World*. Cambridge, Mass.: MIT Press.

Schewe, Phillip F. *The Grid: A Journey through the Heart of Our Electrified World*. Washington, D.C.: Joseph Henry Press, 2007.

Schiffer, Michael Brian. *Taking Charge: The Electric Automobile in America*. Washington, D.C.: Smithsonian Books, 1994.

Schrank, David, and Tim Lomax. *The 2007 Urban Mobility Report*. Texas Transportation Institute, the Texas A&M University System, 2007. http://mobility.tamu.edu.

Seiler, Cotton. *Republic of Drivers: A Cultural History of Automobility in America*. Chicago: University of Chicago Press, 2008.

Shoup, Donald C. *The High Cost of Free Parking*. Chicago: American Planning Association, 2005.

Smil, Vaclav. *Energy: A Beginner's Guide*. Oxford: Oneworld, 2006.

Smil, Vaclav. *Energy in Nature and Society: General Energetics of Complex Systems*. Cambridge, Mass.: MIT Press, 2008.

Southworth, Michael, and Eran Ben-Joseph. *Streets and the Shaping of Towns and Cities*. Washington, D.C.: Island Press, 2003.

Sperling, Daniel, and Deborah Gordon. *Two Billion Cars: Driving toward Sustainability*. New York: Oxford University Press, 2009.

Stern, Nicholas. *Stern Review on the Economics of Climate Change*. London: HMSO, 2006.

Strzelecki, Ryszard, and Grzegorz Benysek, eds. *Power Electronics in Smart Electrical Energy Networks*. London: Springer, 2008.

ULI—the Urban Land Institute and NPA—the National Parking Association. *The Dimensions of Parking*. 4th ed. Washington, D.C.: Urban Land Institute, 2000.

Urry, John. *Mobilities*. Cambridge: Polity, 2007.

Vairani, Franco. BitCar: Design Concept for a Collapsible, Stackable City Car. PhD thesis, Department of Architecture, MIT, June 2009.

Vaitheeswaran, Vijay, and Iain Carson. *Zoom: The Global Race to Fuel the Car of the Future*. New York: Twelve, 2007.

Vanderbilt, Tom. *Traffic: Why We Drive the Way We Do (and What It Says about Us)*. New York: Knopf, 2008.

Weinert, Jonathan, Chaktan Ma, and Christopher Cherry. "The Transition to Electric Bikes in China and Key Reasons for Rapid Growth." *Transportation* 34 (2007): 301–318.

Weiss, Malcolm A., John B. Heywood, Elisabeth M. Drake, Andreas Schafer, and Felix F. AuYeung. *On the Road in 2020: A Life-Cycle Analysis of New Automobile Technologies*. Energy Laboratory Report MIT EL 00-003. Energy Laboratory, MIT. October 2000.

World Bank. *Cities on the Move: A World Bank Urban Transportation Strategy Review*. Washington, D.C.: World Bank Publications, 2003.

Zahavi, Yacov, and Antii Talvitie. "Regularities in Travel Time and Money Expenditures." *Transportation Research Journal* 750 (1980): 13–19.

Illustration Sources

1.1, 1.2 General Motors

2.1 General Motors

2.2 AAA: "Your Driving Costs," 2008

2.3 Clockwise from top left, from published information—Honda FCX Clarity, Tesla roadster, FIAT Phylla, BYD F3DM, Chevrolet Volt, Great Wall Smart EV

2.4–2.6 General Motors

2.7 Carnegie Mellon University, www.tartanracing.org

2.8–2.11 General Motors

2.12 Smart Cities

2.13 General Motors

2.14 Smart Cities

2.15 United Nations Population Division, Department of Economic and Social Affairs

3.1 Urban Congestion Report: National Executive Summary, Federal Highway Administration, April 2007

3.2 Smart Cities

3.3 General Motors

3.4 World Business Council for Sustainable Development: Mobility 2030

3.5 Dinesh Mohan, *Mythologies, Metros, and Future Urban Transport*, Indian Institute of Technology, January 2008

3.6 General Motors

4.1 General Motors

4.2 Clockwise from top left, from published sources—Toyota i-REAL, Suzuki PIXY+SSC, Nissan PIVO-2, Honda PUYO

4.3–4.12 Smart Cities

4.13 Segway

4.14–4.17 General Motors

4.18–4.19 Franco Vairani

4.20–4.22 General Motors

5.1 General Motors

5.2 Argonne National Labs, "The Greenhouse Gases, Regulated Emissions, and Energy Use in Transportation (GREET) Model v. 1.8b," http://www.transportation.anl.gov/modeling_simulation/GREET/

5.3 Energy Information Administration

5.4–5.6 General Motors

6.1 Smart Cities

6.2 McKinsey & Co.

6.3 Smart Cities

6.4 Smart Cities, from information published by Coulomb

6.5, 6.6 Smart Cities

6.7 Smart Cities, from information published by KAIST

6.8 Smart Cities

Chapter 7 (opening image) Franco Vairani

7.1 VENcorp

7.2–7.6 Smart Cities

7.7 John Williams and Smart Cities

8.1 M. D. Meyer, *A Toolbox for Alleviating Traffic Congestion and Enhancing Mobility*, Institute of Transportation Engineers

8.2–8.9 Smart Cities

9.1 United Nations Population Division, Department of Economic and Social Affairs, *World Urbanization Prospects: The 2005 Revision*

9.2 Federal Highway Administration

9.3 General Motors

9.4 Renault

9.5 U.S. data collected from various sources; UNECE 2001 (cars European Union and Accession Countries, population Accession Countries), Eurostat 2001 (population European Union) and WorldBank 2002 (GDP in US$ and 1995 prices Accession Countries and European Union)

9.6 Source data from R. L. Forstall, R. P. Greene, and J. B. Pick, "Which Are the Largest? Why Published Populations for Major World Urban Areas Vary so Greatly," City Futures Conference, University of Illinois at Chicago, July 2004

9.7 Jesse H. Ausubel, Cesare Marchetti and Perrin Meyer: "Toward Green Mobility: The Evolution of Transport," *European Review*, vol. 6, no. 2 (1998), 137–156

9.8 Federal Highway Administration

9.9 Data from Jeffrey R. Newman, Felix B. Laube (eds.), *An International Sourcebook of Automobile Dependence in Cities: 1960–1990* (2002)

9.10 Federal Highway Administration

9.11 General Motors

9.12 Smart Cities, after Don Shoup

9.13 Federal Highway Administration

9.14 "New York Household Travel Patterns: A Comparison Analysis," based on 2001 National Houshold Travel Survey, by Pat S Hu and Tim R Reuscher, Report prepared for the Office of Transportation Policy and Strategy at New York State Department of Transportation, Albany, New York 12232. Prepared by Oak Ridge National Laboratory, Oak Ridge, Tennessee, 37831-6285. Report ORNL-TM-2006/624

9.15, 9.16 General Motors

9.17 Smart Cities

9.18 General Motors

9.19, 9.20 Smart Cities

9. 21 General Motors

9.22, 9.23 Smart Cities

10.1 General Motors and Smart Cities

Index